AF706517

Mathematics and Scripture

Jesus Villalobos

Published by Jesus Villalobos, 2024.

MATHEMATICS AND SCRIPTURE

First edition. September 15, 2024.

ISBN: 979-8224704446

Written by Jesus Villalobos.

Mathematics and Scripture

By

Jesus Villalobos

Chapter One

What Has Math Got To Do with Scripture

Justify your arguments. Any court on earth begins with opening arguments. They take the form of accusations and denials. Why does this seem as if it is a universal theme? This is undoubtedly Scriptures' basis as we read the debate between God and Satan for the soul of Job, which is the earliest book of the Scriptures known.

Understandably then it stands to reason that mathematics also needs to be proven as the basis for Scripture if WE intend to move forward in any direction, equating math with Scripture. This shall be the opening of this book.

You'll notice I used the word WE in the previous paragraph and not the word I. This of course is a subliminal message as I begin this book. For it is not I alone who is involved with writing this book. It is the Lord and I as you will soon discover through the manifold wisdom involved here.

If it were merely I, this book would have never been written. Therefore I must include the Lord in its opening credits. This then, becomes the reason for the word WE.

So just what has math got to do with Scripture? One cannot help but notice the innumerable times certain numbers come up in the Scriptures. The numbers one, two, three, four, five, six, seven, eight, nine, ten, eleven, twelve, fifteen, seventeen, thirty, thirty nine, forty, fifty, and myriad combinations of these are scattered throughout the Scriptures.

Yet although scattered, they seem to hold mysteries as they are used time and time again in patterns and rhythms which boggle the human mind. Can this be but mere coincidence? Or is there a much higher reasoning behind such things, which intends for us to find this symmetry?

It is beyond my intellect to believe that mere coincidence could be behind any such anomaly. Rather, it begs us to consider that the mind behind such things could only be called God if one uses rational thinking.

Therefore, this shall be my opening argument for that with which this book shall deal. For it behooves me, to argue on behalf of my God and His Son Yeshua commonly called Jesus, who is known as the Christ and rightfully so.

With this thought then, let us, (God and I), move forward. There is much to teach, much to enlighten, and much to explain. Let us begin with the word opposites. For the old saying that opposites attract, is one that is commonly known across the globe.

It is almost as if to say that Scripture and mathematics could be considered opposite sides of the same coin so to speak. By this I mean that they are worlds apart, and yet compliment each other as positive and negative, night and day, right and left, or spiritual and physical.

For the mind of God can only be seen by the physical attributes of this creation known as the fabric of space and time. This being the case let us take a closer look at the mind of God through mathematics, which is the measurement of both space, and time?

Why the question mark? It is because although it appears as if both can be measured, modern science is in the throws of discovering that both are illusions. By illusions, I mean that both are merely shadows of a much greater reality, which has already passed by.

For those of you familiar with quantum physics you already have a clue as to what I'm talking about. For those less fortunate who remain ignorant of this sector of science, what I am about to reveal will be both astonishing and eye opening.

But what has quantum physics got to do with Scripture you might ask? Much more than most are capable of either accepting or grasping. Nonetheless, I will try to show their relationship without causing you to think I've gone mad or am suffering from vain delusion.

Science has ascertained that all matter is composed of photons which are particles not of energy, but rather light. Since this is the case, we find that Scripture backs this finding in a couple of places.

Genesis tells us that in the beginning of the spoken Word of God, light is prime or first. Genesis 1:3-5. These Scriptures showing plainly that all things would require light to exist. Far fetched? Let's add another Scripture.

John 9:5 "...I am the light of the world." This Scripture showing that Jesus who also is prime, or the firstborn of the dead, considered Himself to be the light of the world.

Of course Jesus was speaking of Himself as our leader showing us the way of righteousness which was to follow Him in word and deed, crucifying our flesh through His death and resurrection. Nonetheless, He calls Himself our light or our understanding.

I want you to consider for a moment that He used light as His word of explanation for a logical reason. For all things that can be known about God are known through the physical attributes of our known universe. We find this same logic in Romans 1:19-20.

Since Jesus calls Himself the light or our understanding of just who God is, we can safely say that Jesus equated light with understanding and since light is prime or first, it takes understanding or reasoning to know God.

So then reasoning itself becomes prime or first. Logic based on what can be seen becomes our rationale for both believing on and loving our God. But we could take it one step further.

We could also say, that it is God's understanding, which created all things. Since light and understanding are equated by Jesus Himself. Therefore, we could say that Jesus is the manifestation or light of God's thoughts.

The person of Jesus then, held all God's thoughts and showed them to us in increments. Bite sized pieces of light or understanding which showed to us the mind or thought process of God so we could understand Him.

Jesus knew all things pertaining to creation, since He Himself was the essence of all that is made. He is God the Son, and is the light of the world. All things made, exist through Him.

Therefore, all things exist through light or understanding. So then it is only one step further to equating that light is the manifestation of understanding or logic in the physical universe.

If indeed this is the case, then light reveals the intellect or reasoning of God. Hence, the logical order of creation reveals the logical mind or reasoning of God, thus showing His existence and His love.

I am speaking scientifically because mankind is physical and needs physical instruments to measure things by. He discerns by what his five senses can grasp. According to God, this is enough of an explanation or proof that He exists and loves. Romans 1:19-20.

Since God declares this to be a truth, it is only logical, that the proof of God should start with the very prime or elemental fabric which links all things together. This would be light. Genesis 1:3-5.

Therefore, science saying that the visible universe is created out of photons which are the smallest particles of light would not be wrong in its assumptions. But there is a mystery when the universe is viewed this way.

Photons have no mass! This creates an enigma. For in order for the universe to exist, indeed mankind himself, there must be mass. Mass is the amount of space something takes up. If there is no mass, there is no object!

For our conception of reality is one in which all things take up a certain amount of space which gives us the element of time as well. For time cannot be measured outside of a starting point.

Now perhaps you can see sciences dilemma. For if our very fabric of space has no mass, then time also has no measurability since it lacks a starting point to measure it by. Therefore, what is, is. Nothing can be added or subtracted from it. It is, and always has been, complete and unchangeable.

Are these not the attributes of our God? "I am the Lord I change not!" Malachi 3:6. "Jesus Christ, the same yesterday, today, and forever." Hebrews 13:8. Jesus then, has God's attributes within Himself. He is God. He is, and has always been complete within Himself as well, just as His Father is.

But let's get back to science and the study of quantum physics. Photons, from which everything is made, have no mass. Therefore, what we know as creation should not and cannot exist using conventional logic.

Yet everything within us tells us that it does exist. For we experience it with our five senses. What this tells us is that either it exists, or we don't! Now our logical mind tells us that this universe exists. Therefore it would be lunacy to say otherwise?

Yet according to our tools for measuring this universe which are made from the rudimentary elements of our planet such as gold, silver, copper, physics, knowledge of electronics, etc., This universe itself is beyond our measurement capability. This, because photons have no mass. Even when using our minds in combination with these elements, we fall short of a standard measuring facility based upon photons having no mass.

But if we use our minds alone, we can measure the things, which we know as creation. We find that both time and space are measurable. But our minds have created the very fabric of what we call measurements.

For measurements of both time and space are impossible through what is known as science. They have already proven this through their studies of quantum physics. Photons have no mass and since all creation is made up of photons, creation has no mass. It does not exist!

Yet we can discern it with our minds. So which is it? Does it exist or doesn't it? The most logical reasoning will tell us that it exists within our minds! But if that is the case, who put it there?

Since it is evident that mankind cannot even figure out this enigma through his own intellect, it requires a higher intellect to have done it. We call Him God and Father. He it is who is responsible for what we experience on a daily basis. He is responsible for what is known as creation.

So we see then that the Word of God is true when it says that God created everything out of nothing. For if indeed photons have no mass, then all things made with them have no mass. Hence, this creation was made out of nothing!

Now any being that can make something out of nothing should be called God. Hence this is the logic for both believing and giving Him honor. To be otherwise minded would be foolishness. Hence, Romans 1:18-32.

Once again, science is in the midst of a dilemma. If they try to locate a photon's position, they cannot find its speed, if they measure the speed of a photon, they cannot discern its location. It loses its mass.

The most logical conclusion then, would be that time and space are one. One, being merely the reflection of the other. Much like electromagnetism is the opposite of magnetism, or energy is the opposite of matter. Or Scripturally, Jesus being the reflection of His Father.

What we know as time, would be the equivalent of creation, which is the equivalent of what God calls His Word, or which the Church calls Jesus Christ, the Word of God.

What is known as space would be the equivalent of God the Father within whom, are all things pertaining to creation. Hence, Jesus' words concerning Himself being in the Father, and the Father being in Him would be extremely accurate.

Of course I am speaking scientifically, but this is how mankind measures things. God and Jesus are in fact persons, not mere physics. But since mankind is himself a person, he should be able to ascertain that it takes sound reasoning to believe in God. For creation and his study of it, demand he reach this conclusion.

Or what other logical conclusion is there that the universe is made out of nothing? I can find none. You are welcome to keep trying if you wish to befuddle yourself. But God and I know this is foolishness.

This would be the reason for several Scriptures, which appear to be both in the present tense and the past. The chieftest or prime of these Scriptures being Ecclesiastes 3:15; "That which is (exists) now, has already been; and that which will be (will exist in the future), is nothing new (has always existed as well)."

So we see that God is, and nothing can ever change what is, because all else is vanity. It doesn't really exist except in the sense of God creating it using nothing. It cannot challenge God since it is nothing, and God is! He alone exists!

Chapter Two
Probabilities

Chance, otherwise known as probabilities in more eloquent circles, is something that is universally accepted as the deciding factor in what is considered truth verses what is considered falsehood. How many people rely on such things as a simple coin toss, or the luck of the draw in the short straw wins?

But herein lies the difference between rational thinking, and what many call coincidence. For who knows if the tossing of the coin itself decided whether or not the football game was won or lost? Who knows if the life of the person who pulled out the short straw was decided by the straw itself?

Does this sound like foolishness to you? How much more foolish do you think God considers men who ignore words (our Bible) so finely tuned which can transfer a man's destiny from His eternal wrath to His infinite mercy?

Probabilities then become very important in understanding the mind of God. For they are in fact the very measurement of time, space, and fundamental thoughts of both science and religion when viewed correctly.

Not being a mathematician by any means of the word, I will not try to stagger you with calculations, which will in all probability mean nothing to you. Rather, I will appeal to your sense of logic in the area of probabilities.

Since this book deals with mathematics and the mind of God, let us consider a watch, which is made by mankind. The watch itself is unaware that a man made it. This is also the state of mankind in his early years before he is able to use the fabric of time and space to rationalize.

Now the maker of the watch does not want his craftsmanship to go unnoticed. He therefore puts his name or his brand name, on his craft. One might call this vanity on the part of the watches' creator.

Nevertheless, those who encounter this watch can plainly see who made this watch and will accept it as fact. This discounts of course counterfeits. But for the sake of argument, let us say the markings on the watch are legitimate.

Though the watch has no mind of its own, we as humans look for an identifying mark to see who made the watch. In so doing, we decide whether or not it was made by a reputable source and whether or not it is worth purchasing. This because of previous purchases made by the same manufacturer.

Now let's say one is not acquainted with the manufacturer. One must go about the business of asking others who are acquainted with the manufacturer as to its durability and precision. This is much like those who do not know God on a personal level. They must rely on those whom they believe have more knowledge about God than they themselves.

But what of the watchmaker? Surely he wishes those who know of his craftsmanship to spread the word concerning his excellent work and the reliability of his product. Surely he wishes to expand his business, his name, and his reputation. This of course due to vanity on some level.

Am I accusing God of vanity by this comparison? Or course not. However, we can compare the two in the sense that God wishes His craftsmanship known by that which he has made.

The watch has no mind to give credit where credit is due. However, we as human beings do have minds and because of this are required by our maker to give Him credit for that which we have received due to His gracious heart and hands.

Just as the watchmaker puts his mark on his craft, so God puts His mark on that which He has created.

Scripture puts it this way. Ezekiel 9:4-7. God tells the angel to place a mark on the foreheads of those who sigh because of the iniquities (sins) being committed in Jerusalem.

Those upon whom the mark is placed will survive, while those without the mark will be struck dead by the angels whom God has commissioned to slaughter. This mark identifies those men, who are God's.

But there is another type of mark, which exists. It is what is known as a counterfeit or knock off. God acknowledges that this mark exists as well. He is against it in much the same way that any reputable watch manufacturer would be against an inferior product bearing his name or brand mark.

We see this mark first being introduced in Genesis 4:15. Although God places this mark upon Cain Himself, we know that God was against Cain and what he had done to his brother Abel. This we can know by the fact that Cain was banished from his relatives. Genesis 4:9-16.

Although Cain is promised that vengeance will be taken by God if anyone slays him, yet we see that in fact Cain is slain later by one of his own relatives. Genesis 4:17-24. "If Cain is avenged sevenfold, then Lamech seventy times seven."

By the above words, we can deduce that eventually, Lamech would also be slain by one of HIS own relatives at a later date. Using sound wisdom, we can calculate that what Lamech was stating was just the opposite of what Jesus stated when Peter asked Him how many times he should forgive his brother, up to seven times? Jesus stated that Peter should forgive his brother not only up to seven times, but up to seventy times seven. Matthew 18:22.

Here we see something of the mind and or heart of God. We see that Cain's line would perpetuate the murdering of its own line, while the godly line would perpetuate the forgiveness of theirs.

But why does God use the number seven as a basis for His requirement of forgiveness? And why extend it to include the exponential of multiplying it by a number divisible by itself; forty-nine? 7x70=490.

The answer if you can receive it is that God is showing us His law of love. For the Ten Commandments, were originally only seven. They can be subdivided into ten, which is why God shows us both versions, in a play on words here.

For if we multiply seven times ten, we come out to 70. Thus we have not only the original seven commandments being represented by this commandment to love thy brother, but it also includes the sub division of the law into its ten commandments to love.

Thus we see that God's law is one of total forgiveness for His children. As it is written; "...and forgive us our trespasses (crimes against fellow believers in Jesus) as we forgive those (who are our brothers in Christ Jesus) who trespass against us..."

This is a universal law for the children of God in Christ Jesus His only begotten Son. Without this law, the children of Satan could not be distinguished from the children of God.

The mark of Cain placed upon him by God, is intended to differentiate his type of life which is carnal, from the spiritual type of life found in Christ Jesus. One is the mark of a murderer incapable of forgiveness, while the other is the mark of one who loves and is only capable of love which includes by its own nature, the ability to forgive.

Both are equal in stature. Both are marked in Scripture by the same base number; seven. But they are worlds apart and are in fact complete opposites. One steal, kills, and destroys, the other gives, heals, and replicates.

This then gives us our first look into the mind and or heart of the God of the universe. Cain's family line continuously murdering and cursing itself by its own words, and God's family line which continuously nurtures itself by the Word of God's law of forgiveness.

So what we learn then, is that the number seven is indicative of the law of God, which is love, which brings complete rest or tranquillity. Genesis 2:2-3. God is love. He obeys His own commandments, which are by His own nature how He lives. 1st John 4:16.

Now the law of love is one that demands sharing. Otherwise, it is not love, since love that is not shared becomes selfishness. Therefore, it became necessary for God to show love not merely once, but continuously in order that he might exude his very life from age to age.

As it is written; "In order that He might show the exceeding riches of His grace and mercies toward us in Christ Jesus." Ephesians 2:7 and Romans 9:22-23. These Scriptures showing both races. Cain's and Jesus'. Cain's being made for destruction, and Jesus' to be glorified.

These two family lines being one of flesh which is selfish by nature, and one being of the Spirit of God which is loving, which demands sharing by its very nature. This being the reason that the carnal mind is enmity with God's mind. Romans 8:7.

But let us delve into this a little further. This will give you an even better understanding of the number seven throughout the Scriptures. We can find out much if we apply our hearts to Wisdom.

Scripture denotes that there are seven spirits of life contained within God's laws. For the law of God is life itself and its characteristics (spirits), are contained within the law. But there is also the inverse of the seven spirits of life.

These are the seven spirits of death. It was to these seven spirits of death, which Jesus referred, when explaining to the Pharisees their ultimate doom when they were resurrected in their houses (fleshly bodies). Matthew 12:43-45, Luke 11:24-26.

While most Christians see this as Jesus warning the boy out of whom the demon was cast out not to sin again, what is actually happening is Jesus is speaking to the Pharisees concerning their debate with the other Apostles while Jesus was up on the mountain with Peter, James, and John.

The debate was about the validity of Jesus' claims to be the Messiah. The Pharisees had been mocking the disciples because they could not cast the demon out of the boy which in their minds, proved that Jesus' claims to being the Messiah were not valid, seeing his disciples could not cast the demon out in His name.

Therefore, in His anger against the Pharisees for their hard hearts, he tells them their latter end after they have died. This is what He means when he speaks of the evil spirit going out of a man. He is telling them what will happen when their evil unbelieving spirits leave their bodies.

They will go through waterless places seeking rest and find none. Then they will declare to themselves that they will go back to their house from which they came. This is speaking of them being resurrected as the house refers to their physical bodies.

It is at this stage, that they will take upon themselves the seven spirits of death mentioned in Proverbs 6:16. "There are six things God hates; yea seven are an abomination to Him..." The reason the seventh is separated from the other six, is because it refers to the blasphemy against the Holy Spirit which is unforgivable.

The number seven then refers to the law or laws of God. They bring peace or rest when obeyed.

Notice that the seventh day is one of rest for God. He both blesses and sanctifies it. But just what do these two words mean and why does God choose the seventh day to bless and sanctify?

If one is to believe today's analogy of the seventh day used in today's Churches, one states that the number seven denotes completion. This however is not the case, since it was on the sixth day and not the seventh that God completed His creation.

Therefore, it is the number six that denotes completion in the physical sense. I state that it means completion in the physical sense because there is another number that means completion in the spiritual sense. I will discuss that number a little later in this book.

But God does not leave us ignorant of what the word bless means. He is careful to word Scripture in such a way as to give us the meaning of certain words which have been for too long now, misused by today's clergy. They not explaining properly what is being said when they use these words.

Let us first look at the word bless. God uses this word when dealing with Adam after her (Don't be alarmed, for more information on Adam being created female first, read my book; "Time for Truth.") being made.

Genesis 1:28. "And God blessed them saying, "Be fruitful and multiply..." This tells us that to be blessed means to have the ability to multiply. This then is what happened to the seventh day. It was multiplied which means it became more than what it originally was.

So what was one day became at least two days. It may have become more days, but it at least became two days. Thus we see that God extended the seventh day.

But why is this important? It becomes very important to understanding prophecy where Jesus speaks of His thousand-year reign. For it is only after the sixth day, each day equaling a thousand years, that Jesus begins His earthly reign. Although it speaks of a literal thousand year reign, it is understood by the fact that God extended the seventh day, that Jesus' earthly reign will be extended for at least two days, or two thousand years, and is most likely referring to the fact that He will rule eternally, yet upon a newly created earth that will last forever.

The logic behind this is that a day to God (Jesus) is as a thousand years. This according to Psalms 90:4 and 2nd Peter 3:8. This means that the seventh day is extended and is at least two thousand years long, but more accurately is speaking of His rule being eternal. Jesus comes back after the sixth thousandth year is completed to rule upon earth for the seven thousandth year.

To fully explain this would take another book inside this one. You can get the full explanation by purchasing my book; "Three Days of Darkness and the Coming Flood."

Satan is imprisoned for the first thousand years of Jesus' earthly reign, and then released to gather the armies of the world against Jesus. God the Father destroys them with fire and brimstone from Heaven.

Then begins the second day of Jesus' reign on earth. But Scripture doesn't tell us by how much the seventh day was blessed, so we don't know how long it is until the Father destroys Heaven and earth to recreate them for Christ and the Church to live in. Revelation 20:11.

The above Scripture reveals that God the Father will destroy Heaven and earth at some point to recreate them. Jesus spoke of this day in Matthew 24:35-36. Many misuse these Scriptures to prove the timing of Jesus' return is not knowable. This is not what Jesus said. He said the timing of Heaven and earth's destruction by His Father was not knowable.

He expects His own to know exactly when He will return. But as I said in the above paragraph, it is too much to explain in this book. Get my book; "Three Days of Darkness and the Coming Flood.

What is evident, is that there will be people born upon the earth between the time Jesus takes charge of it, and the recreation of Heaven and earth. This may be so that God can fill Heaven with creatures, which his children can rule over. These creatures of course being created in conjunction with the new heavens and earth, but this is merely speculation on my part.

But the population of saved humans must be adequate to handle such things as must be looked after for eternity. Therefore, God allows mankind to repopulate the earth until such numbers are reached to adequately rule over the new Heaven and earth.

Am I positive this is the reason for the blessing of the seventh day? No, but it is a possibility. One that bears mentioning. But for whatever reason, the seventh day was blessed or extended.

Now we come to the other word, which needs to be explained. The word sanctified. To sanctify something means to set it apart for a special purpose. This can be for evil or good. In this case, it is for a godly and good reason. The reason is that it is for the sake of allowing mankind to continue multiplying.

This is evident in that when Jesus returns to rule on earth, there are still people alive in their fleshly bodies who continue to multiply the population of the earth. So then this is at least part of the reason for the multiplying of the seventh day and setting it apart. Since Jesus begins His earthly rule on the seventh day, or seven thousandth year from mankind's being made.

It is also evident that it is during this seventh day (each day being one thousand years) that the actual rest of this world comes into its fullness, with Christ and His followers ruling over this present world. So once again, we see rest associated with the number seven.

Once again I would like to reiterate that it is the law of God or love which brings about this rest to which mankind becomes the beneficiary. To further prove that this is so, I will place into this portion of this book a portion from one of my other books; "Time for Truth," which if you have already read, you can skip over and not miss anything.

This next section will further your understanding as to why God's laws will bring peace to this earth. It will also increase your understanding in just how God intends to bring peace through His Church.

Revelation 1:12 pictures the Church as seven golden candlesticks. Revelation 1:20 confirms this to be true. But the question is why? Why is the Church sevenfold? The answer is given to us in Revelation 4:5. "...and I saw seven lamps of fire burning before the throne (of God the Father) which are the SEVEN SPIRITS OF GOD (the Holy Spirit).

Do you see it? The Holy Spirit is sevenfold just like the Church. This is the other half of God and the image of this half of God is the Church! So what we see then is that in Genesis 1:26, God the Father was not talking to Jesus. He was holding a conversation with His female gender that made Her own image, which we know now as the Church.

Jesus is the image of God the Father, while the Church is the image of the sevenfold Holy Spirit.

Why you might ask would the Holy Spirit or the Church be shown to be sevenfold? We must go back to the original Hebrew language to find the answer to this question. Bear with me as I explain this.

When God gave the law to Moses on Mount Sinai, there were only seven commandments. What? Yes. In the original language, the commandments numbered only seven. They can be subdivided into ten, which is why we call them the Ten Commandments instead of the seven commandments.

What does this have to do with the reason why the Holy Spirit and the Church are pictured as sevenfold? It has very much to do with it. The seven commandments or seven spirits of life which are contained in the laws of God bring eternal life.

Each law would be compared to a spirit or an attitude of God's righteousness. The Church and the Holy Spirit have these seven spirits of righteousness within them.

In other words, the Holy Spirit and the Church obey the sevenfold law of God. One other Scripture which verifies that the Church is indeed the image of the Holy Spirit is Proverbs 9:1. "Wisdom (the Holy Spirit), hath built Her (female gender) house. She (female gender) hath hewn (carved) out her (female gender) SEVEN PILLARS. This is clearly saying that the Holy Spirit has hewn out Her image, which is the Church.

When we look at Proverbs 6:16, we see seven things that God hates. These seven things that God hates are the exact opposites of His divine sevenfold law.

Therefore, since God's laws bring eternal life, the breaking of these seven laws must by interpretation bring eternal death! This is the reason why the Holy Spirit and the Church are pictured as being sevenfold. They have seven SPIRITS OF LIFE!

There are Old and New Testament Scriptures showing different Scriptural pictures of the Church being sevenfold.

Zechariah 3:9 "...upon one stone (Christ), seven eyes (the Church) shall look." Why is the Church shown as eyes? Well, what do you do with your eyes? You witness things. We are Christ's witnesses!

Now, if you look at Zechariah 4:2 "...behold a candlestick all of gold, with a bowl on top of it and seven lamps thereon, and the seven pipes to the seven lamps, which are upon the top thereof:" This Scripture speaks not of the Church, but of the Holy Spirit. This is made clear in Zachariah 4:6 "...not by might nor by power, but by My Spirit say's the Lord of hosts."

This once again shows God's Spirit is sevenfold. Remember that God's laws in the original are sevenfold. The Spirit and the Church carry this law of God within themselves.

Looking to Joshua 6:4 "And seven priests shall bear before the ark seven trumpets of ram's horns." These seven trumpets and priests are meant to be symbolic of the sevenfold Old Testament Church and the angels or leaders of the Church. The leaders symbolically represent the seven Spirits or laws which control the Church. We see this symbolism again in Revelation 1:12.

Again in Revelation 1:16 which speaks of the New Testament Church symbolically calling them stars and angels; Revelation 1:20. Revelation 5:6 "... stood a Lamb as it had been slain, having seven horns (kingdoms or laws), and seven eyes, which are the SEVEN SPIRITS OF GOD sent into all the earth (it is the Church which has been sent into all the earth). The sevenfold Spirit of Christ indwells the Church. They follow God's law!

It took the death of Jesus to accomplish placing the Spirit into the Church. Once He died, God took the female portion of Christ and placed it into His female gender the Church.

This is not blasphemy; this is what the Word of God declares. This completed the image of God the Father and God the Holy Spirit. John 7:38-39 "... out of His (Christ's) belly, (the literal says womb just as the literal of where Eve came out from Adam), shall flow rivers (seven) of living waters

This speaks of the sevenfold Holy Spirit, (but this He spoke of the Spirit, which they that believe on Him should receive: for the Spirit was not yet given because Jesus had not yet been glorified). In other words, Jesus had not yet died and resurrected!

Once the Spirit was given to mankind based upon Jesus' death and resurrection, the Church needed to be built up just as the woman who came out of Adam had to be built up. What this is saying, is that there had to be a learning process in order to bring the woman who symbolized the Church, to maturity. And so it is with the Church. She must learn just what pleases and displeases her husband. The Church is to be a bride. Any bride who is to be worth her salt must know how to please her husband.

Genesis 2:22 "And the rib (the literal says side implying a gender) which the Lord God had taken from Adam's womb Genesis 2:21, built He up into a woman." She became Adam's wife just as the Church will become Christ's wife. Both came out of their respective wombs.

The Church will reign alongside Jesus. This will fulfill the prophecy of Genesis 1:26 "...and let them, (speaking not of Adam and Eve, but rather Christ and the Church), have dominion over the fish of the sea (gentiles), and the fowl of the air (devils), and over all the earth, and over every creeping thing (animals), that creeps upon the face of the earth." Once this dominion or ruler ship is established, peace or rest shall ensue. This is why God's laws bring peace. When obeyed, they bring life and happiness.

Chapter Three
More Numbers

Increasing wisdom does not come at a small price. It comes with countless hours of study and prayer. This has been the case in my life as I began to read the Scriptures at the age of eight. Once I started, I could not stop. Nor have I ever had the desire to stop.

Study began almost immediately after I began to read as I wanted to understand just who God was and what He had done for mankind. I mean this in the sense that He had saved mankind with a purpose in mind.

I decided early on, to make it my mission in life to honor God by understanding all I could, concerning His doings and reasoning. This has been my goal in life as I continue to worship Him for giving me salvation through His only begotten Son Jesus Christ.

To know the mind of God is something which is not taught in today's Churches as much as it ought to be. It is all important if we are to understand, why God has done certain things, and why He will do other things during the course of history.

Wisdom comes in two forms. It comes in the natural and it comes in the spiritual. Both can be used for good and godly reasons. Understanding is learning to put to use, that which we have learned.

To understand the mind of God is one thing. To allow this to change us conforming our lives to the will of God is something that takes years of God working with us, which is why we are allowed to continue living once we have been converted or born again.

God's desire is for us to not only understand Him, but to actually become like Him in deed. He desires to rule over our minds and hearts so that we can have the same peace that He has. But this peace comes through obedience to His sevenfold law of love.

This obedience is something that is learned as we go through this life allowing God to take control of our minds and hearts a little at a time. It is a progressive thing and should not be neglected by us as His children.

Let us take a look at the number one in Scripture shall we? We can find many things out concerning God through this number. It is this number which denotes God who is one.

But the concept of oneness is one that needs to be understood through God's eyes and not our own. Herein lies a truth, which has been too long overlooked by today's leadership.

Elohim is the first word used for God in the Scriptures. It is a plural word. It would not be wrong to interpret this word as meaning "The Gods". Many cringe at this interpretation fearing that it implies more than one God. This is not the case. This is merely the first mention of the one true God. This is the first hint at the trinity. "In the beginning, the Gods created the heavens and the earth." Genesis 1:1.

This then begins our journey as we seek to unravel the picture, which God has laid out before us to understand. Your understanding of who God is will be greatly enhanced if we can get you over this first hurdle in the race to find a relatively simple truth hidden in a rather complex matrix.

Let God explain God and you will be quite surprised at what you will find. Please remember that God intends to give us His entire plan for mankind's creation in these opening passages of Scripture.

God gives us another clue to Himself in Genesis 1:26; "Let Us make mankind in Our image." Once again we see plurality associated with our God. This concurs with the opening Scripture and the use of the word Elohim. But just what is God saying to us? Is He not saying to us since He Himself is speaking in plural form here that His creation of His image must by logical conclusion be dual as well?

God the Father is speaking to someone else other than Himself when He makes this statement concerning creating mankind in their image. The real question is to whom is God the Father talking? If you ask this question to the average leader of today's Church world, they tell you that God was speaking to His Son Jesus.

This however cannot be the case for one very important reason. This statement about creating mankind in God's image is not the end of God's statement. God goes on to say that He intends to give mankind rule over all the earth bound creatures. This statement made by God is not a conditional statement.

Therefore, it cannot be speaking of natural man who is sinful. For God would never give unconditional rule to a sinful creature. Therefore, this unconditional rule which God the Father intended to give was not to natural man, but rather was a prophecy concerning Christ and the Church. This is why there are no conditions attached to this prophecy of giving mankind rule over everything concerning the earth.

If therefore, this statement concerning giving rule over the earth to mankind was the Father speaking to Jesus, it really does not make sense. Since God is telling this other person that they together give their image dominion. This would mean Jesus was going to create His own image and then give it dominion. This is absurd. The most logical answer to the mystery of who God was talking to is that God the Father was talking to the Holy Spirit concerning making Christ Jesus and His bride the Church in the image of God the Father and God the Holy Spirit.

We can see this to be true by many Scriptures, which tell us that the dominion of the earth is to be handed over to Christ and the saints. Daniel 7:22 "...and the saints possessed the kingdom (earth)". Matthew 25:34 "...Inherit the kingdom prepared for you by My Father from the foundation of the earth."

The Scriptures declare Jesus to be the image of the invisible God (the Father). Hebrews 1:2. We also see that the Church is called the bride of Christ.

This is the other half of mankind, which God said would be made in His image. Therefore, it would behoove us to try to understand the Church makes up the second half of the image of God.

If then God was in fact talking to the Holy Spirit when He declared He desired to create His image upon the earth, who constitutes the Holy Spirit's image? This becomes all-important if we are to mature in what God has ordained for us to understand.

God gives us the answer in Genesis 1:27; "So God created mankind in His own image, in the image of God created He him; MALE AND FEMALE created He them." This is not as many believe a simple statement telling us that God created both genders. This is an amazing declaration by God Himself telling us that His image is both male and female. Any other interpretation would be a blatant misrepresentation of what God Himself has declared in His Divine Word.

God is telling us that Jesus is the image of the Father while the Church is the other half of the image of God, which is the female gender. The Church is the image of the Holy Spirit! Why would I say such a thing? Because; it needs to be said. The Scriptures prove this statement to be true.

The oneness of God is a plural oneness which although is separate, remains one. In other words, the same life that flows through God the Father also flows through the Holy Spirit and the Son.

It follows the same logic as when God declares that all races of men were made from one blood. Acts 17:26-27. In other words, each man is his own man, yet he owes his life to the one source from which it came in the beginning; Adam's life.

This is the earthly picture of the spiritual. God the Father is the original source of life, which now resides within the Holy Spirit and His Son. Three persons, having the same type of life within them.

Let me explain something here. Jesus Himself declared that He must rule until all God's enemies were placed under His feet. Then, when this is accomplished, He will hand over the kingdom to His Father.

In other words, the kingdom is not in the Father's hands at this time. So we do see a separation between the Father and the Son. The Son will need to hand over the kingdom to His Father at some later date. 1st Corinthians 15:25-28.

It is at this time that God will once again be all in all. So then the oneness which God speaks of in His nature has to do with the unity between the Father, Son, and Holy Spirit in the sense of a unified purpose. The purpose being the salvation of mankind.

Jesus speaks of the Church being one with Him even as (exactly as) the Father and He are one. John 17:20-21. Does this mean anything like what has been expounded in the Churches today? No, it means that we have the same spirit or type of life that Christ has. This enables us to be unified with Him in purpose and live a godly lifestyle. We are not however the same person. We become like Him in lifestyle because the same type of spirit (a holy or godly one) abides within us!

This is what God calls oneness. It is a unity so strong that it becomes impossible to separate what we are doing from what God the Father, Son, and Holy Spirit are doing. This unity of mind and heart is what God calls oneness. This then is what the number one is indicative of in the Scriptures.

This number is equated with God who is the original source of this type of oneness. He accomplishes this by being the very source of life, which gives itself to reproduction. This is why we are called the children of God. We have the very life of God living within us. Without this life flowing through our human bodies, we have no real life within us since God is and has always been the only true source of life from eternity.

The natural life is only a facsimile of the true life, which flows from and through God the Father. He has shared this life with those whom He calls sons and daughters; those who have been born again. They have received the very Spirit or life of God within their human bodies.

Does this explanation offend you? What shall we say then when we are called sons or daughters of God? Why should one be offended to be called the offspring of God? The alternative is to be the offspring of Satan, or to have nothing but the facsimile of life, which is the earthly spirit, which is devilish. 1st Corinthians 2:12. In this Scripture we see that the world has a spirit (life) to offer, but this life is evil and not godly. This is the type of life the natural man has received to his own destruction.

The natural or carnal mind (spirit) is enmity (at odds) with God. It does not subject or unify itself to God's purposes nor is it able to. Romans 8:7-9. Christ has portioned to us a portion of His own Spirit, which He received from His Father. Thus the very life of Christ flows through our human bodies, in the same way that the life of Adam also flows through our bodies. Its source came from Adam.

In the natural then, we are one in Adam. But in the Spirit we are one in Christ because the life we have received came out of Christ and was portioned to each of God's children making them alive through this life giving Spirit.

One life source, for all who have partaken of this life source. This is what God calls oneness and this is what all who are God's children have received. They are in fact God's offspring. They have the very life of God flowing through their earthly bodies which will be changed to become like Christ's glorified body.

We have explored the number one; let's move on to the number two and its implications. We can still learn much concerning the character of God and His magnificence through this number as well.

For it is this number which speaks of witness or proof of truth. John 8:17-18 and Deuteronomy 19:15.

If the witness of two or more men placed the sentence of death upon a man in the law, how much more shall the testimony of two produce life eternal upon those considered worthy to share in Christ's eternal life?

For the law is become a two edge sword. It both accuses and acquits those who are in Christ Jesus. For the law of love which God has shown through the gift of His only begotten Son shall greatly surpass the letter of the law which was given on Mount Sinai.

The former being witnessed by the law and the prophets. Romans 3:21, Matthew 17:1-3, and Mark 9:2-4. This transfiguration of Jesus was surely a prophetic vision of His soon to be glorified body. Moses and Elijah, being a picture, of the law and prophets, which bore witness of our need for salvation.

Surely in this vision, though it is not specified, Jesus' marks of His crucifixion were evident on His glorified body even as they were evident when He showed them to His Apostles in the locked room after His resurrection. Even though His Apostles still did not understand fully this vision.

This vision then comes with agreement to the two-witness law of God. It embraces this law by showing Christ's death needed two witnesses to establish it as a truth. His death then was truth in the sense that He needed to die for the sin of mankind done in the Garden of Eden by Adam and the woman.

How ironic that the death of two people (Adam and Eve), needed to be healed by the death of two people on the cross. Christ and the Church. Once again we see the number two involved here. This being a witness of God's truths.

Looking at Scripture, we see Jesus sent His disciples two by two to the neighboring towns to preach and to heal. This to let people know that their witness was not just one, but twofold which made it true. Of course they had a third witness which was the healing power of God.

It is a sad day in which we live that we rarely see this twofold witness being used in today's Church. It seems as if the Church of today always seems to think they can do God's instructions one better.

But God says what He means and means what He said. This is one element that has been abandoned as a rule, which needs to be replaced to its proper place in the order of things.

Now you might think me legalistic, but then if we abandon Christ's teachings, we abandon Divine order. This is one and the same as chaos. For we were sent to do God's bidding and not our own. God can only bless the order of things that He placed for us to follow.

One might think along the lines that the Church has grown enormously in the last few decades which shows God has blessed the Church's efforts. However, if we really take a close look at today's headlines, we find that there are more earthquakes, more floods, more tornadoes, more people hungry, and more people sick.

What can I say to these things? God gave us a promise, which is one I cling to. He promised that if those people who are called by His name shall humble themselves and pray and seek His face (His favor by obedience) and turn from their wicked ways (of doing it their way), that He would hear from Heaven and heal the land.

But is this what we see happening? Any logical mind, which can still reason, will tell you what we see is the complete opposite of this healing promised by God. Why? Because the Church is still doing it Her way; instead of God's.

How dare you attack the Church of the living God you say? I dare to say what God has put into my mouth to say. How can you go on in your ignorance of believing a lie?

Today's preachers would have you believe that the Church is better now by far than she has ever been. If this were the case, why are there fewer miracles, less people living Godly lives, less people satisfied with their Christian walk?

It is all good and well to attend Church services and feel good as you sing unto the Lord praising Him. But what about your home life? Your children are in rebellion, your spouse is at your throat and refuses to speak kindly to you, and you feel lost and without direction in your spiritual walk.

If these things aren't true I am a liar and the truth is not in me. But you know in your heart that these things are true. You put on your Christian face and smile while all the while you wonder why you don't feel the closeness promised to you by your Pastors.

Good! This is the beginning of your healing. Healing comes with doing what you know to be the right thing. Scripture puts it this way. "To him who knows to do good and does it not, to him it is sin."

Loving God is more than words. It is doing what you know to be right. It is dying to your own selfish desires and living for the truths of God's words, which include doing what He says.

If the Church as a whole were doing what God instructed her to be doing, things would not be what they are in this world. We would see much more healing, more miracles, more spiritual direction, and more love.

What am I getting at? Let me give you an example. We go to Church and sing a few songs. Afterwards we sit and listen to a bunch of words, which a Pastor is speaking on behalf of God (hopefully).

The person on your right hasn't eaten in two or three days, but you are unaware of it as you get your spiritual fix. The person on your left needs his car fixed but cannot afford to take it to a mechanic.

You are an expert mechanic, but you are unaware of his needs, so he leaves the service with the exact same problem he came into the service with. He still has a broken car and no way to fix it.

Now the Pastor feels he has done God's will. He has spoken God's message. He has fulfilled the needs of His sheep. WRONG! Christ put it this way; James 2:15-16, "If a brother or sister be naked, and destitute of daily food, and one of you say unto them, 'Be ye warmed and filled," what doth it profit, seeing you give him not the things which he needs? Even so, faith without works is dead."

Yes, he is using something physical to relate something spiritual, but the point is, that what he is using in the natural is what shows the spiritual condition of a man. Love nurtures that which is lacking.

A Pastor's job is more than to feed people spiritually. His job is to see that his congregation's physical needs are met as well. But how can this be accomplished when they use the word fellowship to mean a gathering that disallows its members to present their needs openly to one another?

Am I alone in my witness; or is not the Word of God with me in what I have written? The early Church used to gather together for a full meal, which was followed by the Communion Service.

It was at these meals, that one member could discuss their needs with those who sat next to them. Now God, in His wisdom, would make sure that the proper people sat next to a person in need.

He would allow a person who could fill the needy person's need to sit next to him. If he could not himself meet the need, he knew someone in the congregation who could. This is what the Scriptures meant by fellowship.

If one were in need of spiritual guidance on a particular issue, God would seat a person who had been through something similar who had come through it and could give godly advice on how to handle the situation.

But alas, the Church has gone her own way and abandoned these early traditions. Why? That is the big question. Why abandon something that worked, for some pseudo type of fellowship?

Was it because it lessened the work a Pastor had to do? Was it due to inconvenience? One can only guess. But I CAN tell you that true fellowship is more than a gathering to hear one man speak.

People read their Bibles and are content for the crumbs of God's blessings, when they could be enjoying the three course meals that God longs to give them. But how can He bless something He never ordained?

He gave us a pattern to follow. We see it early in the book of acts. They knew each other's needs and met them. How do you suppose they knew each other's needs? They discussed them at the dinner table of course.

Now I wish to be fair in my judgments. I have seen some of this type of fellowship in very small increments practiced in the Church. But it is not by far a common practice, it is done only occasionally, and many times with a charge to its congregation.

This is far from what the Apostles ordained in their examples written in the Scriptures. For too long now there has been an ignoring of what has been written for our benefit and instruction. How long will you abandon what God has placed in order?

How long must the children of the highest God continually suffer due to your refusal to fall in line with Scripture? Do you suppose that God is pleased by this abandonment of His truths?

The answers to your congregation's needs are right in front of your eyes. How long will you keep them shut? How many years must go by with your sheep suffering due to lack of a true shepherd who cares for them as his own children given to him by God to care for?

Do you not see Jesus feeding the crowds? Do you not see Jesus sending out His shepherds to heal the sick, raise the dead, and bring salvation in all its forms both physical and spiritual?

Many in your congregations long to speak of these things, but say nothing due to their feelings of inadequacy. They feel as if they would be scorned for paying too much attention to the physical needs.

Let me tell you something. Jesus created us physical beings with needs. He expects us to help each other with those needs. But if you deny your congregations the ability to communicate their needs, you do them a disservice. These things ought not to be so.

Do you suppose oh Pastor who is in charge of multitudes that you by yourself are able to fulfill the needs of thousands? This is not what God intended. Let us take a look at what happened to Moses.

Exodus 18:13-27. For the sake of brevity, I will simply explain what happened. Moses needed help to judge the people's cases. His father in law Jethro, advised him to appoint judges over thousands, hundreds, and fifties. Moses paid attention to sound advice and things went well.

Each member of a congregation has talents, which are being wasted. They themselves know full well that their talents could be used to a much greater degree, which would glorify God.

But they are restrained because they rarely hear of any needs of the congregation which fall along their special talents. So they go through their lives only occasionally being able to use those special talents, which God gave to them to fill their brother's needs.

This is largely due to the lack of sound leadership. As I have said, how long will you ignore the examples set forth in Scripture? Many needs go on for months or years due to this lack of sound leadership, which denies its congregational members to exercise their God given talents.

Instead, they produce a pseudo type of fellowship, which sends their congregational members home still cold and empty, their physical needs remaining just as they were before they gathered for this pseudo fellowship.

Don't expect an apology from me for speaking truth. If you are cut to the heart consider it the circumcision lacking in your life as one of the leaders who is guilty of this very thing.

Do not suppose I am saying I am any better than those reading this book. I have my own sins in need of correction. My life is not perfect by any means of the word. Yet if I remain silent and say nothing which God has ordained me to speak, I become guilty before God for withholding words which could avert His judgments which are already falling upon this world.

Once again I reiterate, the natural disasters falling upon this world are God's judgments due to His people's lack of showing love towards one another. To alleviate this, God's people must begin showing true love for one another which means to meet each other's needs.

This can only begin happening as the leaders begin to take action to ensure true fellowship between believers. They need to contemplate ways to assure that their congregational members are able to relate their needs to one another on a regular basis and not merely single out one member once in awhile during a particular Church service!

What will it take for the leaders of the Church to understand what is the will and the heart of God? Did not Jesus say that the world would know we belonged to Him by our love one for another?

But love is more than a "God bless you," upon entering a Church service. It is a lifelong commitment of caring for each other's needs. For the love of God can only be seen through our physical actions!

Is this not true even in the physical world? How do you show love for your wife or children? With words only? Of course this is foolishness. Yet by thinking words from a pulpit satisfy physical needs which could be satisfied by simply allowing true fellowship to occur in your service is also foolishness!

Let the body take care of itself just as your own body heals itself. The body cares for itself if allowed to work properly. By all means, speak the word of God, but allow those who are able to meet the needs of their neighbor to do what they are able to do for them.

Allow part of your service to be the body taking care of the body's needs. Is this so much to ask? Yet this is a major reason why the world as a whole speaks against the Church. They see a pseudo love.

What words can I use to make you understand a basic truth, which God has placed into His Word for you to follow? Love thy neighbor. By so doing, you show the love of God within your heart. For how can the love of God be in you when you deny the basic needs of every man woman and child?

You send donations half way around the world to aid in disaster relief, while those in your own congregation go hungry and lack suitable clothing to wear. This due to your lack of leadership skills which would allow the body to take care of itself.

They sit right next to the one they are supposed to help, and are denied the chance of hearing his or her need. How ridiculous this appears to God who sits on His throne daily waiting for a wise man among you.

Do you suppose God is pleased with this type of operation? When a man needs his belly filled to be at peace and all he receives is words? When a woman lacks clothing for her children and all she receives are words?

The answers to their needs are right next to them during the Church service. Why do you need to be struck with harsh words for you to understand these things? But you will say, "We have a food and clothing operation in our Church for our members.

Okay, good. But how many members are even aware of it? How often do you as a leader mention it? And what of the one who has other needs? The member who desperately needs a mechanic to work on his or her car so they can continue going to work?

What about the one who needs spiritual guidance concerning an issue ongoing in his or her life? Are you so blind that you cannot see that we as a body have the members within to solve and heal these dilemmas?

How can God bless with spiritual blessings, His children who continually ignore the physical needs of their own congregational body? You wonder where the miracles of old have gone? First take care of what you have power to do, and then God will perform the miracles He is famous for.

But if you cannot even do that which is physical and within your own power, do not expect God to step in and perform the wonders He is able to do. First prove your love for each other, and God will believe your love for Him.

What has all this got to do with the number two? One person caring for another's needs, equals proof of living in true love. There remains much to say, and how am I straightened until it is said.

Matthew 18:19-20. "Where two or more are gathered together in My name, there am I in their midst." This Scripture has been abused beyond recognition. People have been led to believe that it takes two people ganging up on God to persuade Him to do what He already knows to be right.

This is once again ridiculous. The above Scripture was the end of a much longer discourse by Jesus. The full discourse is Matthew 18:15-20. Jesus is explaining what to do in the case of a brother who is willfully being disobedient to the law of love, which is supposed to exist within the Church.

He is quoting from the Old Testament law of appointing judges to hear each man's complaints against his or her neighbor. He tells them to bring the case before the judges for a ruling, and He (Jesus), will be present to carry out the sentence on the offending party.

I confess that I could not find this verse although I used the concordance to find it. However, you may know where it is. I searched for it but could not find it even though I know I have read it before.

This is what Jesus was speaking about when He spoke of two or three gathering in His name. He was speaking of gathering for a judgment against an unrepentant brother. He would be there to carry out whatever sentence was required for the offense.

We see this sort of thing happening when Peter speaks to Ananias and Saphira concerning their deceptive gift unto God. Jesus stood with Peter and the others to carry out the death sentence to both of these persons who tested the Spirit of God.

It is imperative that we as believers understand Jesus' words correctly if we expect Him to answer our prayers. We are under a contract of marriage it is true, but our contract has to be engaged correctly by proper understanding of it.

If we misuse Jesus' words, we cannot expect answers to our prayers. We must follow protocol. We must speak truth and pray according to Jesus' promises and not according to improper usage of what He said.

Chapter Four
What about the Number Three

Three is a number that represents family. It stands for the Trinity. Some deny the doctrine of the Trinity. They are convinced that God is one and only one. They do themselves an injustice in thinking this way. They miss out on much the Word of God wants to reveal to them.

Let's start with the word God. What does the word God actually mean? It is not God's name. It is a title. It is much like the word king or queen or duke or any other title we give to men. This title tells us our rank or station in life.

But the word God does not mean a single solitary being as some claim. The title of God includes three persons. Jesus said to baptize people in the name (not names as though the three persons he said to include in this baptism ritual were separate beings having separate names) of the Father, Son, and the Holy Spirit.

The fact that Jesus used the word name and not names, shows that these three individuals have the same family name! They are in fact one family. Now I know that many Christians will balk at this interpretation, so I wish to explain exactly what I mean.

Genesis 2:24 "Therefore shall a man (singular) leave his father and his mother (singular usage of parents), and shall cleave unto his wife (singular usage showing only one wife), and the two (showing only one couple) shall become one flesh."

This Scripture is a prophetic Scripture. It is supposed to show Christ leaving His Father and Mother which are in Heaven, and cleaving unto His wife the Church. Now if you think I have lost my mind, look at another Scripture, which will verify what I have just written.

Ephesians 5:30-32 "For we are members of His body, of His flesh, and of His bones. 'For this cause shall a man (Jesus) leave His Father (God) and His Mother (the Holy Spirit), and shall be joined unto His wife (the Church), and the two shall be one flesh. This mystery is great, but I speak of Christ and the Church!"

This Scripture is showing that Jesus is part of a family. He left His Father and Mother to be married to the Church. Does this interpretation offend you? Why? Why should the truth be offensive to you?

The Scripture saying a man would leave his father and mother is singular so that it can be understood that it is speaking of a particular man leaving his particular parents. This can only be Jesus, since Adam did not leave his parents to marry Eve. Adam had no earthly parents!

Furthermore, it cannot be speaking of Jesus leaving Mary and Joseph, since Joseph had already died before Jesus had even began His earthly ministry. Therefore Jesus could only have left His earthly mother at this time.

The only possible meaning of this Scripture is that Jesus left His Father and Mother from Heaven. This would be God the Father and God the Holy Spirit. Hence the Trinity; the Divine Family.

One name for three members. It is the same as the fact that each member of an earthly family shares the family last name. Though they are separate beings, they have one family last name. This is why Jesus said to baptize those who believed into the name (not names) of the Father, Son, and Holy Spirit.

Now a person, who becomes born again, becomes a part of this same family. He or she receives the same life flowing through their body. They inherit God's genetic make up. He gives them His actual life. They are quite literally His children. They therefore, now carry His family name.

If you doubt what I am saying, tell me, what is the meaning of the Scripture, which states that all that are God's are given His name. Ephesians 3:14-15 "For this cause I bow my knees unto the Father of our Lord Jesus Christ, of whom (by whom) the WHOLE FAMILY IN HEAVEN IS NAMED.

This is not saying God gave you a new name in the sense of a new first name. It is saying He gave you His name. You now have God's family name written upon your foreheads.

God has shown His nature by His very creations. He has shown us through many examples. He called each of the earthly families through their tribunal name (singular). There are thirteen tribes in Israel. When God speaks to them, he calls them by their tribal name (singular) though they are many people.

Why has this eluded so many people's understanding for so long? It is not a hard concept to understand. But you will say, give me more proof. Very well, let's look at a couple more Scriptures.

Proverbs 6:32-35 "But whosoever committeth adultery with a woman lacketh understanding: he that doeth it destroyeth his own soul. A wound and dishonor shall he get; and his reproach shall not be wiped away (he shall never be forgiven).

'For Jealousy is the rage of a man (singular as in Christ); therefore He (Christ) will not spare in the day of vengeance. He will not regard any ransom; neither will he rest content, though thou givest many gifts.

The above Scripture is speaking of Christ's return. He will slay those who are responsible for drawing His wife the Church into adultery. They shall never be forgiven, even if they offer many gifts.

This type of jealousy and rage is also seen in one other place in the Scriptures. It is seen when Jesus tells the Pharisees that all manner of sin shall be forgiven men wheresoever they may blaspheme. But blasphemy against the Holy Spirit shall never be forgiven neither in this life, nor in the life to come. Matthew 12:31-32, and Mark 3:28-29.

This is not speaking of Christ's wife, it is speaking of God the Father's wife, or Jesus' true Mother whom he left to cleave unto His wife the Church! There is no atonement for this sin.

Now those of you who think the Holy Spirit is merely a force or a power that God possesses had better ask yourself a very important question. What is more important, to say something against a particular talent or strength a person has, or to say something against the man himself.

If the Holy Spirit was merely a power that God possessed, He should not take offense if people blaspheme one or more of His powers above blaspheming Him. This would be as ridiculous as a man getting more upset over the fact that you told him he can't sing worth beans than saying that he himself was a so and so.

No, God is not saying that if you blaspheme His power, you have no forgiveness. He is saying that if you blaspheme His wife, you have no forgiveness. For if jealousy is the rage of a man (Christ). How much more a holy God (Christ's Father)?

Now you will say that to think God has a wife is blasphemy. I tell you to get your carnal mind out of the gutter. By whatever means, God has a female portion, which is called Wisdom or the Holy Spirit. This is akin to Christ's wife the Church which is why both are sevenfold. See Revelation 1:20 and Revelation 4:5 which show both are sevenfold. The Church is the image of the Holy Spirit.

This is not a misinterpretation on my part, but rather a full explanation of what God is saying so you can understand Him better. The Scriptures are quite clear on this matter if you will permit yourself to be educated in them.

So the number three speaks of family. It speaks of the Trinity. One Scripture this reminds me of is Ecclesiastes 4:12 "And if one prevail against him; (Satan prevailed against Jesus while he was a man), two shall withstand him; (Satan was withstood by Jesus' Father God, and Mother, the Holy Spirit), and a threefold chord; (the family unit), is not easily broken.

Now you will have to judge as to whether I am speaking according to proper understanding in the above Scripture. Jesus Himself said that the Scriptures spoke of Him.

John 5:39 "Ye search the Scriptures; for in them ye think ye have eternal life; and these are those which testify of Me." Again; Psalm

40:7 and Hebrews 10:7 "In the volume (the entirety) of the book it is written of Me, I have come to do Thy will (Father and Mother).

If the entirety of God's Word was written concerning Jesus, why do you fault me for using the Scripture in Ecclesiastes, to support that it is speaking of Satan withstanding Jesus, and sending Him to the cross without just cause; and Jesus' parents, stepping in to withstand what Satan had done to their Son?

This is the only possible correct interpretation for this Scripture. To use another interpretation would not go with Jesus' statement saying that the entire Word of God had to do with Him.

Since the Church has no power to withstand Satan alone, once Jesus had been crucified, they could not help Him. No, it had to be His Father and Mother who withstood Satan and the evil he had done to their Son.

So once again, three equals family in the Scriptures. It validates the Trinity. To be otherwise minded is foolishness and does not fit Scripture. This then is the meaning of the number three. But there is another truth having to do with the number three that I would like to show you.

Science and Religion have long disputed over when life actually begins. Does it begin exactly when a sperm enters the egg, or does it begin sometime later after the visible creature shows human extensions?

Do you really suppose that God in His Wisdom would leave us ignorant of when life actually begins? After all, He has a big stake in human life. No, God made it abundantly clear when life begins.

Science knows that when a sperm enters into the egg, it explodes into a million pieces. It then reassembles into what is known as the human DNA. This is akin to what happened to Jesus in the grave. He died and was buried. It took three days for Him to complete the process of bringing life back into His dead body.

By this one act, God forever told us how long it takes for life to regenerate. It takes three full days for life to begin. The exploded sperm takes three days to fully complete the rearranging of its parts to form the human DNA.

Once this process is complete, it is a human life. Thus God answered even this question for us so there should have been no confusion as to when life actually begins. It takes three complete days.

Why do men look everywhere else but the Scriptures to answer hard questions? Instead they turn to science and witchcraft. God has given us the answers if we would only search the Scriptures diligently with our hearts open to hear what saith the Spirit of God.

Why do the heathen rage and set themselves in array against the Lord and His anointed? How long shall they think themselves wise and become fools in the things of God? How long shall they replace godly wisdom with human foolishness? Do they think they shall escape the judgment of God?

They use every excuse under the sun to find a reason why killing unborn life is excusable. They claim that since there is no set time in the Scriptures as to when life actually begins, they can make up their own minds as to when life begins and when it is to be called a human life.

What an atrocity! What arrogance! By choosing to remain ignorant of the things clearly written for our instruction, they have blinded themselves to understanding that they are committing murder, the penalty for which is death!

And what of Christians who side with these monsters? Do they suppose that they are any less guilty of the crime? They remain ignorant of the things of God and pray that they are making the right choice rather than studying to show themselves approved.

Why am I being so harsh? Because there are countless Christians who have succumbed to this foolishness called science to avoid the embarrassment of an unwanted pregnancy. Countless souls have been murdered by professing Christians!

What kind of light, do you suppose this puts on God? When those who are called by His name continually refuse to know Him properly? Making idols in their minds of whom He is, instead of studying Scripture to know Him as He intended.

Do you suppose God is pleased with such things? Do you really think that He is pleased with these makeshift idols that His people have created in their own minds of Him? Am I saying that those who have done this horrible crime of abortion are beyond repentance?

No, but why should they have done this crime in the first place out of willful ignorance? For the answers to life's biggest questions ARE in the Scriptures for us to find if we search diligently with a pure and honest heart.

The problem is that God's children have become lax in their study of God's Word. They refuse to read regularly the Word of God. This is why they remain ignorant and have so many problems in life with major decisions.

They don't know which road to take because they haven't studied the map of God's Word. Instead they choose to remain simple minded babes in Christ. This is nothing more than laziness and I will not say God is pleased with it.

They make many foolish decisions, which end up affecting more lives than their own. Furthermore, many who might have been spared heartache, will now have to go through it due to God's children's lack of wisdom brought on through lack of study.

Their errors in judgment would never have occurred if they had only studied more diligently to find the answers before moving ahead with their plans. But because of willful laziness, they affect many lives negatively and close many doors that might otherwise have been opened for them to show the love of God.

Yes, my words are harsh. But it is only because sometimes truth stings in order to correct. I don't say these things to dishearten you, but to bring you back to sane thinking. I speak to those who are refusing to study.

Those who have had an abortion know full well what I am talking about. They know the pain and anguish and guilt they felt, and many still feel. But it is because they did not study and had no one to guide them into the truth of when life begins.

What shall I say to these things? Shall I praise the Churches' leadership in this thing? I dare say I will not. By their lack of guidance in this area (I speak to those who have not addressed this issue of abortion by teaching when life begins Scripturally), they have opened the door to the murder of unborn children.

Chapter Five
Number Four

The number four speaks of the world. It denotes things pertaining to this present world and its spirit, which is carnal or devilish. Hence it actually has to do with the flesh. When spoken of in the Scriptures, it is always used in conjunction with this present world and never the spiritual world. Let's look at some Scriptures, which will bear this out.

Ezekiel 37:9 "Then said He unto me, 'Prophesy, son of man, and say to the wind, "Thus saith the Lord God; Come from the four winds, O breath, and breathe upon these slain, that they may live."

God explains to Ezekiel that the bones, which He is causing to live, are the whole house of Israel. He tells Ezekiel that He shall bring them up out of their graves. Now this prophecy is not speaking of Israel becoming a nation again as many have led you to believe.

This prophecy is speaking of the resurrection of the just which will be given the land of Israel as their inheritance once God calls them out of their graves. As I keep saying in all my books, all prophecy is literal.

Since God speaks of causing them to come up out of their graves, he is speaking of the resurrection of the just that shall inherit the Promised Land. This and only this interpretation is the correct way to look at these Scriptures.

Ezekiel 37:13 "And ye shall know that I am the Lord, when I have opened your graves, O my people (saved ones), and brought you out of your graves (at the resurrection of the just).

Thus we see a prophecy having to do with the resurrection and each person receiving back their physical body. This is caused, by Jesus giving them a new body. But notice that there are four steps in this process.

- Ezekiel 37:6 "And I will lay sinews (bone holding tissue such as ligaments or tendons) upon you,
- And I will bring up flesh (muscle tissue) upon you,
- And cover you with skin,

- And put breath in you, and you shall live; and you shall know that I am the Lord."

These are the four steps to creating human life. We see these steps again in Job 10:11 except in reverse order.

- Thou hast clothed me with skin
- And flesh (muscle),
- And hast fenced me with bones

- And sinews (ligaments or tendons)."

So we see that the flesh has four steps to its creation. We also see four winds of the earth being used to bring this flesh to life. So once again we see the number four having to do with this present physical world and its creation.

If we look to Revelation 7:1, we see there are four angels at the four corners of the world, holding back the four winds of the earth that they blow not upon the earth. This is so that the winds of the earth will not blow upon the oceans causing them to cover the earth in another worldwide flood just yet. See Luke 21:25-26, which speaks of the sea and the roaring of its waves and it being associated with the powers of the heavens being shaken. This falls in line with Isaiah 13:12-13.

What am I talking about? Notice that the previous verses of Revelation 6:12-17, have to do with the great day of God's wrath upon this world. The sun, moon, and stars have gone dark and it is said that the stars of heaven fell to the earth.

This is speaking of the time both Isaiah and Jesus spoke about in Isaiah 13 and Matthew 24 respectively. Both chapters are speaking of the same event, which is a worldwide nuclear war which will send the earth out of its orbit so that the stars of heaven to those upon the earth will appear to be falling to the earth. But Isaiah 13:12-13 explains to us, that it is the earth leaving its orbit from around the sun.

So the four angels holding back the four winds are doing so due to this nuclear war relayed to us by Isaiah 13, which sent the earth out of its orbit from around the sun. If they were not to hold back the winds, the earth would immediately be covered with the waters of the oceans.

It is at this critical juncture, after the earth has left its orbit from around the sun, that the sign of the Son of Man occurs which Jesus spoke about in Matthew 24:29-31. This sign is three days of total darkness upon the earth as it reaches a place of total darkness in space.

These three days of darkness correlate to the time that Jesus (the light of the world) lay in the tomb. The light of the world went missing for three days as Jesus lay in the tomb. Of course you will say this was a spiritual darkness, and you would be right.

However, we see this darkness elsewhere in Scripture being a physical darkness, which correlates to Jesus' time spent in the grave. We see it during the opening Scriptures of Genesis 1:14, when God tells us that the sun, moon, and stars were made to give light upon the world on the fourth day. Up until then, the earth was still in darkness. See Genesis 1:2.

This darkness, shown to us in Genesis 1:14, is what Jesus referred to as His sign. It lasted for three days upon the earth before God put the sun, moon, and stars to light the earth. This correlated to Jesus' time He would spend in the grave and the light of the world would be missing.

Yes, God had said, "Let there be light...," long before the fourth day, but God explains that He separated or put a barrier between the light and the darkness. He did not allow the light to shine upon the earth at all. Not even on its opposite side. The earth had moved away from its previous location of being near a sun. This is why God explains to us in His Word that the reason He placed the sun, moon, and stars in the heavens was to shed light upon the earth.

But this was not until the fourth day. If God speaking light into existence which He did long before the fourth day had been enough, he would never have created the sun, moon and stars to light the earth.

No, God had merely created the light somewhere else and not allowed it to light the earth at all. The earth was still in darkness until the fourth day. This is why God tells us He separated the light from the darkness. He put a barrier between them.

This is why the Spirit of God is seen hovering over the waters of the earth. It had been flooded. This is just what God will do again just after the rapture of the saints who had died before His Second Coming (those who are alive when He returns go into the millennium after the separation of the sheep and the goats). He will once again flood the earth. This is why He said it would be as the days of Noah when Jesus returned. The earth will be flooded. The only difference being that this time God will not kill all flesh due to His promise to Noah. He will leave ten percent alive.

We see the sign of the Son of Man (three days of darkness) again, when God makes his covenant with Abram, in Genesis 15:12, "And as the sun was going down, behold a horrible darkness came upon Abram." Again we see this sign of three days of darkness, when God delivers Abraham's children out of Egypt; the ninth plague, again when Jesus hangs on the cross; although it is shortened to three hours.

This darkness that falls upon the earth after a worldwide nuclear war (Isaiah 13:12-13), lasts for three days, then Jesus raptures those who died before He returned. His people which are alive are in the mountains out of harm's way as He in His anger turns the world upside down to flood it, just as He did in the days of Noah. Isaiah 24:1.

This is why the angels must hold back the wind from blowing upon the earth. If these winds were allowed to blow, they would cause the flood to happen too soon. They must hold back the flood until the 144,000 Jews who are left upon the earth after the rapture of those who had died before Christ returned, are sealed with God's name have time to join Christians who are in the mountains and have been spared from the worldwide nuclear war's radiation and will be spared the flood being too high up to be affected by it.

They will have been sent to mount Zion to avoid being overtaken by the flood which will engulf the whole world. These men will be the millennium priests. This is why they are spared. Other Church members will be in separate mountains.

They will survive this worldwide flood because they are on Mount Zion, which will be too high for the waters of the worldwide flood to cover. God keeping His promise to Noah not to kill all flesh in the next worldwide flood which was to come later. The rest of the survivors will have gone underground to avoid the nuclear war, which preceded this worldwide flood. These would be non-Christians who didn't flee to the mountains as ordered to by Christ, as well as heathen.

If those who have gone underground to avoid the nuclear bombs come up too soon afterwards, they will be destroyed in the worldwide flood. This is the meaning of Isaiah 24:18-21.

God never said He would not flood the entire world again. He said He wouldn't kill all flesh the next time he flooded the whole world. We can see Isaiah prophesy this next worldwide flood in Isaiah 24:18-19. Notice that the windows of heaven are opened again just as in the days of Noah.

For a more detailed explanation on these events, get my book; "Three Days of Darkness and the Coming Flood."

You will notice that the saints who have come out of the Great Tribulation are shown just before the 144,000 Jews are sealed. These saints are spoken of as having come out of great tribulation. They have endured the tribulation period. They were not raptured before it began. They are saints who died during the tribulation period. So much for a pre-tribulation rapture view. But there is something else. These shown before the throne are only those who had died before Jesus returned to earth. Those alive when Jesus returns, go into the millennium after the separation of the sheep and the goats.

You will notice that God says these will never again hunger or thirst or be smitten by the sun. This is how they died during the Tribulation and it says they could not be numbered. This is NOT saying that there were so many they couldn't be counted. It is saying they refused to be numbered or marked with the number of the beast during their earthly lives.

All these events are said to occur at the end of the tribulation period. Thus we see that the Church must endure the tribulation period due to her laxness in following God. She will have to be purified through much tribulation.

The 144,000 Jews sealed, will become priests in the millennium period to make the sacrifices, which will be given during this time frame. This is why we see reference to sacrifices being performed during the millennium period. These will occur during the feasts in the millennium, which depict Christ's death and resurrection. Zechariah 14:17-21.

We see the number four when dealing with end time scenarios where the four horsemen of the apocalypse are concerned. Each horseman represents a different plague or set of plagues upon the Four Corners of the world. This shows that these plagues are happening upon this present world.

The first horse is white showing the rise of antichrist's false peace. The second horse is red, showing war. The third horse is black showing famine and disease. The fourth is a sickly green, showing death and destruction.

By these four judgments, God will destroy those who have mocked Him as Lord. He will use the earthly tools of destruction within His grasp to render to each according to their own foolishness.

The number four is seen in Ezekiel's vision of the Cherubim. Ezekiel 1:5-6. The creature is shown as having four faces and four wings. This creature is not four creatures in the sense that it has four bodies. It is one creature with four faces and four wings.

Only when we see this to be true can the vision be properly interpreted. This is a picture of the Old Testament Church. For a more complete explanation of this vision, I invite you to get my book: "Time for Truth."

The set of four wings shows four thousand years of Old Testament worship and praise, while the four faces represent the four aspects of the gospel which only the Church can fully preach since they are the ones saved by Christ's sacrifice and not the angels.

Since the four thousand years of Old Testament Church history were only a picture of what was actually going to happen, it can only be the shadow of the truth of Christ's sufferings on our behalf. But the Church must use the Old Testament Scriptures to fully understand all God has to tell her concerning herself and her mission.

And what is the Church's mission? Is it not fourfold as well? Mark 12:33 "To love the Lord thy God with all your heart, soul, mind, and strength. These four make up the totality of the human condition.

Hebrews 4:12 says it this way; "For the Word of God is quick and powerful, sharper than any two edge sword, piercing even to the dividing of soul and spirit, and of the joints and marrow, and is a discerner of the thoughts and intents of the heart.

So we see the four parts of the human anatomy.

- Soul
- Spirit
- Joints and Marrow
- Intents or thoughts of the heart

These are the four parts, which make up the human being's physical characteristics. This shows us the worldly aspects of the human condition. The soul is the emotional and mental, the spirit gives it life, the joints hold the bones together, and the marrow gives the bones their life since it contains the blood cells, and they all work according to a man's thoughts.

The Word of God is very specific when dealing with the human anatomy. It gives a full description of how God put the human body together. God does not leave us without understanding as to how He put us together.

We live in a physical world and have physical needs. We, like the world, have four cornerstones of our creation. Once again, we see the number four has to do with this physical world.

Genesis 1:16-19, tells us God made two great lights to rule the day and night of this world, and the evening and the morning were the fourth day. So we see the fourth day being used to describe this earth and the sun and moon ruling over the day and night.

I could go on and on, but with the above examples, I think you can see why the number four represents the world. So I'll leave it at that and we can continue on with our lessons in what numbers mean in the Scriptures.

Chapter Six
Number Five

As we move higher up in the number system, your understanding will be increased dramatically. Scripture will become very clear as you learn the mathematics of Scripture.

All things created have a mathematical basis. This includes language. Math is the foundation of all things. Math is the universal language. This is why Scripture is filled with numbers. God is actually showing us the formula He used in creation.

Of course He has only revealed a small portion of the formula to us. Jesus said there were things that the Father knew that He did not know. Jesus knows all things pertaining to this present creation or age. His Father did not tell Him some things, which do not pertain to this age of creation.

This is the reason why Jesus said that the times and seasons were in the Father's hands, when His Apostles asked Him if it was at this time that He would restore the kingdom to Israel. Acts 1:6-7.

Jesus was speaking of the time when God would give the new heavens and earth to Him and His bride the Church. Revelation 21:1-3. Matthew 24:35-36 explain that the day when God the Father will destroy Heaven and earth to recreate them in their glorified state is a day known only to God the Father.

We see this happening in Revelation 20:11. The Father reveals His face, at which time Heaven and earth pass away with fervent heat. For our God is a consuming Fire. Deuteronomy 4:24 and Hebrews 12:29.

But what of the number five? What does it represent? The number five represents government. In the parable of the ten virgins, we see that five are wise and five are foolish. This represents the two types of governments within the Church.

You'll notice that the five wise virgins (representing the wise governors of the Church) took along extra olive oil (representative of the Holy Spirit). These were ready when the shout came about the bridegroom (Jesus) coming. Matthew 25:6-9.

The foolish virgins (leaders of the Church) asked the wise to give them of their oil. But the wise virgins (leaders) told them to go to those who sold and bought the Word of God. While they went out to buy, the bridegroom came and they were not ready.

This is symbolic of those who preach that Jesus will come sooner than what He Himself promised, which will be after the tribulation period. Matthew24:29-31. These verses tell us Jesus will come after the tribulation period and not before. But this is not a very common view these days.

The foolish virgins go out to buy, but by the time they have bought the oil, Jesus has already come. These foolish virgins thought they had enough oil because they expected Him to come before the tribulation period began.

Once they realize that they will have to go through the great tribulation, they ask the wise virgins (leaders) to give them from their oil reserve. But the wise virgins only mock them.

What they tell them, is to go to those who sell the oil and buy some (symbolic of the Scriptures, which were written by godly men). This is symbolic that they will tell them to restudy the Scriptures for themselves.

These foolish leaders will begin to study the Scriptures again, but will come to understanding too late. Jesus will have already come by the time they fully comprehend that they should have listened to Jesus' words about His coming being immediately after the tribulation period. Matthew 24:29-31.

The reason the wise leaders refuse to share their oil, is because they know that these men were not true leaders. If they were, they would have known that it wasn't until the end of the tribulation period that Jesus would come.

They preach an escape from tribulation. Jesus told us that if they persecuted Him, they would persecute us. We are not better than Jesus to avoid this type of torture or pain.

I reiterate that the main point of this parable of the ten virgins is about the leadership types. This is not saying that the rapture will occur quickly as they preach. It is saying the exact opposite. They will need extra oil to withstand things until Jesus comes.

The parable of the ten virgins is not a parable unto itself. It is rather, a concluding statement, concerning the fact that there will be servants or leaders of the Church, who are misleading their congregations for personal gain.

They themselves actually believe that Jesus will be coming sooner than what He Himself declared in Matthew 24:29-31. But this delusion is because they are not right in their hearts toward God. Therefore, they ignore Jesus' words in favor of fantasies.

The parable of the ten virgins concludes Matthew 24:37 through 25:1-12. You see, Jesus will only reveal the truth to those whom He knows personally. This is why these other five virgins had believed the lie, which said that Jesus would come before the tribulation period began.

For it was the same in the days of Noah. Only eight people of untold millions survived the flood. Noah and his family believed what God had warned them about. He knew the exact day of when the flood would begin.

Noah knew that he had exactly 120 years from the date God had issued the command to build the ark. It would be on the anniversary of the commission to build the ark that God would flood the world 120 years later.

This is exactly what Jesus is telling those who are listening to Him in Matthew 24:37 through Matthew 25:1-12. He is telling them that they should know when He is coming. But Jesus is also mocking those who will not know when He is coming. He compares them to those in the days of Noah who did not know the day nor the hour of His coming. Only those to whom it was revealed (Noah and his family), knew the day and the hour to be ready to go into the ark (a type and shadow of the salvation of the rapture of the dead in Christ, and the beginning of the millennium for those still alive when He returns).

So then the five wise virgins represent the wise five-fold ministry which preaches when Jesus will come correctly. He will come at the end of the tribulation period just as He promised in Matthew 24:29-31.

Jesus compares the five foolish virgins to what He had spoken of earlier. He said that if that EVIL SERVANT (notice he calls them evil because they expected him sooner) should say in his heart that his Lord (Jesus) delays his coming; this means they expected him (Jesus) sooner. Matthew 24:48.

Jesus promises to come in a day when that servant looks for Him not. In other words, these servants will still not understand when He is coming. Since He delayed His coming in their eyes. But it isn't that He delayed His coming. It is that they misused His words.

They misinterpreted what Jesus said. They chose to follow cunningly devised fables as to the timing of His return. Jesus was plain about when He would come. He is not late as they have said in their hearts. Jesus said He would come immediately AFTER not BEFORE the tribulation period. Matthew 24:29-31.

These evil leaders will have been the cause of countless souls not understanding Jesus' words. They will be appointed their portion with the hypocrites; there will be weeping and gnashing of teeth. Matthew 24:50-51.

The righteous leaders have the extra oil, because they knew Jesus promised to come after the tribulation period, and prepared their hearts to serve him even through the time known as the Great Tribulation. This is why they had the extra oil in the first place. They knew they would need it.

But I still have not shown you why the number five means government. To do so, I must take you back to Aaron and his four sons. Numbers 3:9-10. These Scriptures speak of Aaron and his four sons ruling over the Levites or priesthood of God.

How do I know that Aaron had four sons? Numbers 3:2 "And these are the names of Aaron's sons; Nadab the firstborn, and Abihu, Eleazar and Ithamar. Nadab and Abihu died before the Lord when they offered strange fire (symbolic of strange worship) before the Lord in Sinai (symbolic of the law of God). This is symbolic that they were offering things forbidden by God as a means of worship. These men were leaders of the priesthood, yet they offered strange worship. They were not following God's laws.

Notice that it is exactly half of the leadership which offered strange worship before God. This is exactly what Jesus spoke about in the parable of ten virgins. Five wise, five foolish. Exactly half. This means the Church will be divided into two groups.

One group will understand that Jesus will not come until the end of the tribulation. The other group will follow strange doctrine or leadership. They will believe that Jesus is coming sooner than He Himself declared in Matthew 24:29-31; immediately AFTER the tribulation period.

Yet the leaders of the Levites numbered five. Aaron and his four sons. This is indicative that there are five offices of authority in the Church. They are; the Apostle, Prophet, Evangelist, Pastor and Teacher. Ephesians 4:11.

While most argue that it doesn't matter when you think that Jesus will return, Jesus was specific about those who would think that He was delaying His return. He said that they would most likely begin to beat their fellow servants and to turn to getting drunk. Matthew 24:48-51. Jesus promised to come on a day when they were not looking for Him. This would be AFTER the tribulation.

This would occur due their not understanding that He would not come until after the tribulation period was over. So then we see that in reality, it matters a great deal when you believe Jesus will actually return.

For if you believe what the foolish leadership teaches about Christ coming before the tribulation period begins, you will not have prepared yourself for the hardships which will come during the tribulation period.

You will instead, accuse Jesus of being a liar. This because you believed through erroneous teaching, that Jesus should already have come before the anti-Christ showed up. You will most likely lose all hope. You may call Jesus a liar, when it was those who led you who are the liars.

They chose to follow cunningly devised fables. They have misled you through promises of deliverance from hardship. But serving God was never promised as something that would be easy. Jesus said, "If they have persecuted Me, they will persecute you. If they have called the Master of the house Beelzebub, how much more them of His household?" John 15:20, and Matthew 10:25.

For these know not how to listen to God's instruction. Neither can they follow His instructions. They are like King Saul who made the sacrifice early due to the people's crying in his ears. He didn't wait until evening like he was supposed to. Instead he took matters into his own hands.

When Samuel shows up (a type of Jesus), he rebukes Saul and the kingdom is torn from his hands. Saul loses everything he sought to gain because he loved his leadership more than he loved God. 1st Samuel 13:6-14, and 1st Samuel 15:24-26.

In the above Scriptures, we see a type and shadow of the return of Jesus. We see the people crying to their leadership, we see them crying because of hardship and we see them hiding in the caves and thickets. 1st Samuel 13:6.

This is reminiscent of Revelation 6:15-17. Everyone hides from the wrath of God In the caves and in the rocks of the mountains, because the great day of God's wrath has come.

These men under King Saul should have trusted in the sacrifice, which Samuel would make at the appointed time. Notice they waited for seven days in these Scriptures. But Samuel had still not come.

These seven days represent the seven year period of anti-Christ's rule over the earth and the people awaiting the promised return of Jesus of whom Samuel is a type in this scenario.

The king is a type and shadow of the leadership of those who are awaiting Jesus' return. They will still think they can have control over the people by lying to them and deceiving them as to the appointed time of their sacrifices' (Jesus') return.

But Samuel (who typifies Christ) becomes very angry. This is a type and shadow of Jesus' response to these types of leaders. He tears the kingdom from them. They lose their hope of salvation, which they sought to gain, because they taught false doctrine or offered strange fire (worship) before the Lord. They taught the doctrines of men instead of the doctrine of Christ; due to fear they would lose their authority over their congregations if they told them the truth!

This is exactly what we see going on in the Church world today. The leaders teach an erroneous doctrine that Jesus will return before the tribulation period, when Jesus Himself said it would be immediately AFTER the tribulation that He would return and send His angels to gather the elect saints who were not deceived by this pre-trib view.

These false shepherds water down the fiery words of Jesus because they offend most men. Guess what? Let the words offend them! Jesus' words are Jesus' words! Stop appeasing men when we need to be appeasing God! It is Him whom this world has offended!

So we see that although the leadership or government of the Church is five-fold, it is only half of them that are preaching truth. This is a sad commentary on the condition of the Church, as we know it.

Now one might question the validity of the fact that half of the Church will be misled. I wish to validate what I have written. The fifth office of Apostle is no longer in the Church.

When Paul writes of the five-fold ministry, the Apostles were still around. When they left, they left us with their writings which we call the Bible or Word of God. This became the successor to them.

This is why I say that the four sons of Aaron represent half of the Church leadership. They represent the other four remaining offices, but through the death of two of them, it represents that half of the leadership will not follow sound doctrine, or proper worship.

The word worship is equated by Jesus Himself as obedience to God's laws or instructions. Jesus said, "In vain do they worship Me; teaching for doctrine the commandments of men (over the Commandments or instructions of God). Matthew 15:3-9 and Mark 7:6-13.

The above Scriptures are cleverly used to show a Christian can eat whatever he wants to eat. However, this is not the point of what Jesus said. He said that the way to worship God is by obeying His Commandments or teachings. This includes when He will return and acknowledging that the Church will have to suffer much before He returns due to her laxness of following His Commandments and teachings.

As I said earlier, I have much to say and how I am straightened until it is said as it ought to be said in plain words so even a child might understand what I mean. For the Word of God constrains me to say these things. Let us move on to the next chapter shall we?

Chapter Seven
Number Six

I wrote briefly, about the number six being the number of completion in the physical world. This chapter shall expound on this to show that this is fact. I would not be doing the reader justice if I neglected to prove what I am writing scripturally.

The most obvious Scripture as I wrote earlier is that God finished or completed His work on the sixth day. Genesis 1:31 through 2:1. These Scriptures verify that God finished His work on the sixth day. Thus we see plainly, that six is the number of completion.

However, as with all things, we desire at least two witnesses to establish every truth found in Scripture. Therefore, let us continue on with Scriptures that prove this to be true.

Ezekiel sees a creature, which he calls the cherubim. Now I have heard many interpretations of what this creature is supposed to be, but they fall far short of Biblical interpretation of what this creature represents. I will include an excerpt from one of my other books; "Time for Truth. This will explain what this creature is and why Ezekiel saw it.

There can be no greater honor other than being called a friend to God, than to be given the honor of proclaiming His Divine Word on His behalf. It is a terrible thing to distort truth for personal gain or for vanity. May God grant me favor to avoid such things. As the time of Christ's return draws nearer, the lies and misrepresentation of God's Divine Word seem to be increasing. This should cause a righteous anger to seethe within us and make us be even more diligent in our study of God's Word.

Let us push forward therefore with fortitude in our discovery of truth shall we? I know God will be well pleased with this. And we will benefit much from it as well.

Ezekiel saw a vision of God. It is called a vision for a reason. Anytime the Word of God tells us something is a vision, it needs to be interpreted. If, it were meant to be taken literally, it would not be called a vision but would rather be told as prophecy. This is a cardinal rule when reading Scripture.

Contrary to other visions in the Word, the interpretation is not given to us here in Ezekiel. There is good reason for this. Some things need to be hidden in order to exercise our spiritual minds. This is the idea here with this vision of Ezekiel's. We need to search the Word to find the interpretations for the symbols.

Having studied the Word since a child of eight, with much prayer I hear from the Lord often concerning His hidden manna of the Word. This is one of those times when I really got excited about the interpretation of a vision which has been hidden for so long not only from me, but from the Church at large.

Keeping symbolic language consistent is very important in interpreting God's Word. If we remember this fact, we will be blessed indeed in our studies. God loves to reveal secrets hidden in His Word to His children.

Many are the times I have discovered through the Spirit of God the interpretation of hard to understand mysteries in God's Word. This is a real privilege, which God has allowed in my life. I will be forever grateful for this. I have found out many things about the character and love of God due to the ability, which God has granted me in understanding hard mysteries. This takes discipline and much study, which only God can give to a man.

Understanding Ezekiel's vision is impossible without understanding one key point; the creature in this vision is not four separate creatures, as many believe.

The creature is singular and it is only the faces, which are plural. It is the language, which should have told scholars this.

The language goes from the plural to the singular all through the vision. If we were dealing with more than one creature, the language would always remain plural. This however, is not the case.

Point in case; "Their (plural) form (singular) was that of a man (singular)." Proper English for this would demand that it read as follows if we were dealing with four separate creatures. "Their form was that of men (plural)." It is these types of inconsistencies, which tell us that it is only referred to in the plural due to its faces, and not due to it actually being four separate creatures.

If we can understand this one all-important aspect of this vision, we can understand the vision and what it is meant to represent. We as God's children must come to maturity, if we are going to understand all God has for us. Let us go forward now to fully understand Ezekiel's vision.

Do not be as many who try to understand God's Word without using God's Word as their template. Only God's Word can explain God's Word. The symbols in Ezekiel's vision are within the Word, all we have to do is follow their trail and find the similar traits. Symbols always remain the same throughout the Scriptures. This is why we can use them over and over again to understand visions, which are not explained.

Ezekiel 1:1 "...the heavens were opened and I saw visions (plural) of God (singular)." Ezekiel 1:4 "...behold a whirlwind came out of the North, a great cloud and a fire enfolding upon itself, and a brightness was about it, and out of the midst thereof as color of amber, out of the midst of the fire."

The direction of North implies it came from Heaven. Heaven is in the North according to the Scriptures.

Isaiah 14:12-13 "...I will sit among the congregation in the sides of the North."

These were Satan's words against God. Satan understood that Heaven was in the direction of North. The whirlwind speaks of the Holy Spirit. We find this same analogy when Elijah is taken to Heaven by a whirlwind. 2nd kings 2:11. The chariot, which parted Elijah from Elisha, is the same chariot Ezekiel sees in his vision here. Again we see this whirlwind when Job is visited by God to be reproved for his harsh words. Job 38:1. It is this whirlwind which is also the cloud Ezekiel is seeing.

The fire enfolding itself, speaks of the Father who is self sufficient. Deuteronomy 4:24 and Hebrews 12:29 tell us our God is a consuming fire.

The color of amber, speaks of gold. This becomes important for us to understand that this is the dividing line of that which is in Heaven and that, which is still on earth. This creature which Ezekiel sees is said to have the color of amber above its head. This tells us that this creature is not in Heaven, but on the earth. If this creature were in Heaven, it would be standing upon this glass.

This is also a major key in understanding just what Ezekiel is being shown in symbolic form. This glass is the floor of Heaven.

In Ezekiel's vision, the glass has the color of amber because the creature is below its floor. This is the importance of it being above the creature's head. However, when this creature is seen again in Revelation, we see the glass has become clear and the creature is standing on the glass and is no longer under the glassy sea.

This is because the book of Revelation was given after the resurrection of Christ. The penalty for our sin had been paid for. Ezekiel 1:22 and Revelation 4:6.

"Out of the midst of the fire came also the likeness of four living creatures." This refers only to the faces and not saying there were four separate creatures. "This was their (plural) appearance (singular)." Do you see the inconsistency? "They (plural) had the appearance of a man (singular)," not men as in more than one creature being spoken about.

Ezekiel 1:6 "And every one (of the faces) had (or shared) four faces." This is simply saying that the lion, calf, eagle, and man, was one unit. "And every one (of the faces) had four wings." This is saying that this creature has four wings attached to a human form with four faces.

Now that we understand this creature is only one creature, which happens to have four faces, we can look closely at it to determine just what this creature represents and why Ezekiel was given this vision. Once you understand what Ezekiel saw, you will understand many other Scriptures, which include this creature. Bear with me in this explanation; I promise you will be glad you did. It will enhance your understanding of God's word immensely. Okay, here we go.

Ezekiel 1:7 "And their (the four faces') feet (legs, only two) were straight feet (legs)."

This implies that this creature has no kneecaps! This creature cannot bow down (to man), this creature serves only God! "And the soles of their feet were like calves hooves." This implies split hooves, which is a prerequisite for clean animals. Leviticus 11:2-3 and Leviticus 20:25.

The flesh being like bronze indicates this creature has been tried by fire and found worthy.

Ezekiel 1:8 "And they (plural) had the hands of a man (singular not men, which would make it plural). Having the hands of a man implies that this creature is human! It implies that human hands do its work. This creature then is THE CHURCH!

I realize this flies in the face of what you have been taught concerning this vision of Ezekiel's, but just bear with me a little longer and you will fully understand. I will in fact prove the above statement.

Ezekiel 1:9 "Their wings (four) were joined together; they (the four faces) turned not as they went; they (the four faces) went every one straight forward."

The wings' being joined together, speak of the fellowship of believers. Let me explain. It is the wings, which lift up the throne of God, which is above this creature. The Church lifts up the Lord by its worship and praise.

Therefore, this worship and praise of God joins us together. This is why the wings being joined shows fellowship between believers. We must remember that these wings lift up the throne above it.

The fact that the creature turns not as it goes, speaks of the fact that this creature turns not towards the left, nor to the right of the command of God, but rather, accomplishes God's will. We find this straightforward motion in Proverbs 4:27 and Isaiah 30:21.

Ezekiel 1:10 "...they four (faces) had the likeness (singular) of a man; and the face of a lion on the right side; the face of an ox on the left side; ...the face of an eagle."

These four faces face the four directions S.E.W.N. It is interesting to note that this spells sewn as if a seed (Jesus) has been sown here. The left side speaks of judgment, while the right side speaks of power. This is why the ox is on the left. It speaks of the red heifer, which was sacrificed every year for the sins of the nation Israel.

The lion is on the right because Jesus, the lion of the tribe of Judah, is now seated on the right hand of God in power and majesty. The man's face shows Jesus' humanity, while the eagle to the North, shows Jesus came from Heaven and was and is God. The eagle speaks of Deity or being the supreme judge!

These four faces declare the gospel, which only the Church can fully understand since the angels were not saved through the death of Jesus Christ. Therefore, this creature cannot be an angelic being. This creature is in fact THE CHURCH!

Ezekiel 1:11 "...and their wings stretched upward; two joined one to another, and two covered their (plural) body (Singular)." This shows that our worship and praise of Jesus is directed upward to God and that it also covers our nakedness or sinfulness. Nakedness is the symbol for sin according to the Scriptures, which is why the wings cover the body of this creature. Jesus died naked. Jesus became sin according to the Scriptures. 2nd Corinthians 5:21.

This is also why Adam and Eve were naked in the Garden of Eden; their nakedness was symbolic of their true nature, which they did not understand until the first law handed down by God. It was only after the law came that they understood their sinful nature. Adam and Eve did not fall into sin; they were created sinful! God did this so they could learn the difference between good and evil.

Genesis 3:7 and Hebrews 12:2 and John 19:23 and Matthew 27:35. These verses tell us Jesus died naked, but also, that Adam and Eve's eyes were opened after they ate of the forbidden fruit. This means their understanding of what sin was became complete.

Ezekiel 1:12 "And they (the faces) went every one straight forward (obedience); wither the Spirit (Holy Spirit) was to go, they went; and they turned not as they went (Obedience)."

Ezekiel 1:13 "As for their (plural) appearance (singular), it was like burning coals of fire, and like the appearance of lamps." Revelation 1:12-13. "...and the fire was bright, and out of the fire went forth lightening."

Ezekiel 1:14 "And the living creatures (speaking of the four faces in plurality), ran to and fro as the appearance of lightning." This creature shines just as Jesus does in His glorified flesh, she is His Old Testament bride who runs to and fro with the gospel.

Ezekiel 1:15 "Now as I beheld the living creatures (faces), behold one (singular) wheel upon the earth by the living creatures (faces) with His (singular) four faces."

As we can see, the word HIS refers to the creature as one single creature and not four separate creatures. This should be proof positive enough for anyone! Also we see only one wheel mentioned here, so any future reference to this wheel even in the plural form should be understood from this vantage point. When the word wheels, is used, it is speaking of the fact that this one wheel has another wheel inside itself.

Ezekiel 1:16 "The appearance of the wheels (plural because one is inside the other) and their workmanship was like the color of beryl (a hard greenish mineral);"this speaks of life. The color for life is green.

This is why a rainbow of emerald (green) surrounds the throne of God. Revelation 4:3.

"...they four (faces) had one (singular) likeness (of the wheel): and their workmanship was as it were a wheel in the middle of a wheel."

Now the word for wheel is the same exact word for Heaven (Gilgal). Therefore we might put this another way. We might say there was a Heaven in the middle of a Heaven. This is what God is in fact trying to tell us by this vision. Jesus said it this way "I am in My Father, and My Father is in Me." John 10:30and John 14:10-11. The wheels being green symbolize that life is in the Father and Son.

Ezekiel 1:17 "When they (the faces) went, they went on their four sides (directions) and they turned not as they went (obedience)."

Ezekiel 1:18 "As for their rings (speaking of the wheels), they were so high that they were dreadful."

This speaks of Heaven's sphere of power; "and their rings were full of eyes round about them (the four faces)."

The word full here implies a definite number of eyes not that the wheel was stuffed with eyes. It implies nothing lacking or completeness.

The number of eyes in each wheel was seven (this number represents rest). Since there were two wheels, there were fourteen eyes total. Seven is the number for rest since God rested on the seventh day. However, these seven eyes speak also of witnesses. They speak of the fact that this creature which is the Church, witnesses of heavenly things upon the earth. The outer set of seven eyes symbolizes the (Holy Spirit) after whom the Church is modeled.

The inner set of seven eyes represent the Church which witnesses of Jesus as the Son of God. Zechariah 3:9 speaks of the Holy Spirit while Revelation 5:6 speaks of the Church which mirrors the Holy Spirit, which is also spoken of as sevenfold just as the Church, is in Revelation. Revelation 1:20 and Revelation 4:5.

Ezekiel 1:19 "And when the living creatures went, the wheels went with them."

This shows God follows the Church with signs and wonders to verify He sent them.

"And when the living creatures (the Church) was lifted up (figurative of the resurrection and rapture of the church after the millennial reign), the wheels were lifted up (God goes with the church)."

Ezekiel 1:20 "...and the wheels (representative of the Father and Son) were lifted up over and against them (the faces or the Church); for the spirit (life) of the living creatures (the Church) was in the wheels (the Father and the Son)!"

Ezekiel 1:22 "And the likeness of the firmament upon (over) the heads (four faces) of the living creature (singular), was as the color of the terrible (awesome) crystal (speaks of pure gold which looks like glass), stretched over their (the faces') heads above."

The glass is above the head of the Old Testament Church because Jesus had not yet died to give them access to Heaven. Later, in Revelation, we see this creature standing ON the sea of glass. This however, does not mean that the whole church is already in Heaven, but that God sees us this way since the penalty has now been paid by Jesus His only begotten Son. Revelation 4:6-8.

The creature is the same one shown in Ezekiel but it now has six wings instead of four and has received its fully mature name; "Seraphim." The added two wings are because two thousand years have passed since the completion of the Old Testament covenant. Each generation of worship and praise (which the wings represent) was going to be equal to one of God's days which according to the Scriptures, is one thousand years. Psalm 90:4 and 2nd Peter 3:8. Thus we see that God is praised for every day of the week. Six thousand years equals one week and thus will begin the day of our rest in Christ (the millennium).

Ezekiel 1:23 "...their wings one toward another (fellowship through worship and praise which lifts up the throne of God). Each (face) had two (wings) which covered their front and backside of their bodies (covered their nakedness symbolic of their sins covered because of our worship of the Son of God and His Father)."

The word bodies, is understood in light of the fact these four creatures share the same body, and individually claim the body as their own.

Ezekiel 1:24 "...I heard the noise of their wings like the noise of great waters; (speaks of nations and tongues and tribes); Revelation 17:15. As the voice of the Almighty, (the Church echoes the words of God), the voice of speech (the Word of God), as the noise of a host (army): when they stood (when the Church was silent), they let down their wings (worship and praise)."

Ezekiel 1:25 "And there was a voice (Jesus') from the firmament over their heads (Heaven), when they stood (were not active), and let down their wings." God giving the Church her new orders.

Ezekiel 1:26 "And above the firmament that was over their heads; the likeness of a throne (in Heaven) as the appearance of a sapphire (blue which speaks of loyalty) stone. And upon the likeness of a throne, one sat as the appearance of a man (Jesus) above (the throne) on it." This explains the vision of Ezekiel for you.

There is one more thing I would like to tell you. Satan is said to have been the anointed Cherub. This means he held the same office as the Church. He was supposed to be a priest. Ezekiel's vision was the Church, she is called a Cherubim. Satan then held the same office. We are priests unto God. Since the Church is female, it would seem likely that Satan is also female.

This would explain why Eve, was not frightened by Satan in the Garden of Eden. She was talking to another female! The word serpent is only a reference to Satan's character, not meant to be taken literally.

But we can see here also (due to the creature eventually having six wings), that the Church would go through six thousand years in order to reach maturity or completeness. This then is a second witness that the number six equals completion in this present physical world.

There is much to learn in the following chapter, so let us press forward to the high calling of God in Christ Jesus. You have learned much, but there remains much to learn. Come my children, God desires your growth to maturity.

Chapter Eight
Number Eight

Looking at the number eight, we can see it means new beginnings. One might also say it means new creation, for it is after the millennial reign of Christ or after the seventh day in God's eyes, that He destroys Heaven and earth to recreate them.

Thus we could call the new creation an eighth day. God the Father sees fit to destroy Heaven and earth in order that a permanent Heaven and earth can be created. This is where Christ and His bride will rule.

This will fulfill the prophecies of Genesis 1:26 giving mankind (Christ and the Church), dominion over the earth and its fullness. This prophecy had to do with Jesus and the Church, not Adam and Eve.

To prove this, all one has to do is to notice that there are no stipulations upon this dominion. This means God has every confidence in those to whom He is giving dominion over the earth. God could not have this type of confidence in sinful man.

God speaks of the eighth day when He is speaking of the feasts. He tells Israel that they shall hold a solemn assembly on the eighth day. Number 29:35 and 2nd Chronicles 7:9. Now one might be inclined to believe that this is simply talking about the eighth day of sacrificing unto the Lord, but God knows how many days are in a week, and He did not say, "And the following week on the first day of the week, you shall sacrifice..."

This shows that God knows about an eighth day of creation, which will be an eternal day. It shall never end, nor shall any of its creations end. This is the new creation God will hand over to Christ and the Church to rule over. The Father destroys this present temporary creation, then recreates it to last forever.

We see this happening in Revelation 20:11 and Revelation 21:1-7. Here we see what God will create for Jesus and His bride to rule over. They shall rule forever in this new creation. There will be no end of it at all.

I would like to reveal something at this point, which has been sadly overlooked by today's preachers concerning the book of Revelation. When speaking of the bride of Christ, Revelation shows her coming down from Heaven as a bride adorned for her husband.

This is where people seem to have left their spiritual minds at the door. They fail to realize that the city, which is being shown to John the Apostle, is not a city in the literal sense. We can be sure of this, because it is fifteen miles square and high, and yet only has one street. Revelation 21:16,21 "...and the street (singular) of the city (the city being a picture of the Church), pure gold, as it (singular) were fine gold (There is only one way, one path, and it represents Christ).

The city is the Church and Christ is the way. Notice that the city has twelve gates and twelve angels, which have the names of the Old Testament children of Israel. Revelation 21:12. Notice also, that there are twelve foundations, which uphold the walls of this city, on which are written the twelve names of the twelve Apostles of the Lamb.

The reason why these stones, which John sees have the names of the twelve patriarchs of the Old Testament tribes of Israel, and the New Testament Apostles, is because John is being shown the Church in picture form.

No one can see the Church as she really is in all her glory, except for Christ. She is His and His alone. No one can see her as she is except for her husband. Likewise, no one can see Christ as He really is, except for the Church. She alone has this privilege.

Another place in the Scriptures which shows the number eight means new beginnings, is the story of Noah and the flood. We find that Noah took along with himself seven other people aboard the ark. Genesis 7:7.

Eight people in all were used to start a new beginning upon the earth. From these eight people came all those who are on the earth today. The old earth had passed away, and a new earth had begun. This of course typifies Christ and the sevenfold Church beginning the newly created Heaven and earth.

Many believe the number eight shows salvation. In a sense it does, but it more means new creation, or new beginning. There is another number, which means salvation. I'll talk about it a little later in this book.

It really is the same number as eight, but it is a two digit number, which when added together, equals eight, It is a two digit number because it shows a truth which once again is hidden in symbolic form.

I have actually already spoken of this truth, but it is hidden in a numerical symbol for us to find as well. But enough about this, I have much to teach you if you are willing to learn.

Since this book is both about Scripture and Mathematics, I would like to show you something. The number eight is written in our culture somewhat like a pretzel. It twists and goes around reconnecting in two equal, yet opposite circles. One circle overlapping the other at the center.

I mentioned before about the eighth day being a new creation that lasts forever. If you know anything about mathematics and its symbols, you know that electricity is often shown as a number eight.

This being symbolic of it being eternal. I find it interesting that science has many symbols, which originated with Scripture as their basis. I do not wish to get into too much detail about this; I just think it is very curious.

The new symbol for what they are calling the god factor is similar to the way Hebrew spells out God's name; three upward strokes with a line underneath them. But most of you have probably never seen this symbol, unless you are involved with quantum physics. Since the typewriter keyboard lacks this symbol at present, I insert the closest thing to it here; "∈". Turn this symbol so the three prongs are facing upward and square the bottom so it is straight instead of rounded.

This would be what the symbol looks like for the god factor. It is almost comical how God makes them choose these symbols for their discoveries or postulates. They are speaking truth in symbolic form without even knowing it.

As the Scripture puts it; "Always learning and never able to come to the truth." 2nd Timothy 3:7. They willfully reject the knowledge of God. They prefer their intellect to God, so they may become gods unto themselves.

This of course is foolishness. But they are blind and cannot see. They live in pride and arrogance. Their end shall be the pit. Their own pride and arrogance shall be their destruction.

It is with this attitude that they have devised the very means by which God shall destroy them. It is written, "Their silver and gold shall cling to their flesh and burn them." James 5:3. This is speaking of the nuclear weapons that they have spent their money to build in order to rule the earth. This is why James makes it clear that he is speaking of the last days in this Scripture.

James is not unaware of Isaiah chapter thirteen and other Scriptures, which speak of the day of the Lord as being a day when nuclear weapons are used to destroy the ungodly. Of course James didn't know the term nuclear weapons, but he understood it would be by some kind of fire that God would destroy ungodly men based upon Isaiah thirteen and other Scriptures.

Also, the Holy Spirit caused James to use this connection between the silver and gold of men, which they love above God and the use of it which would spell out their final end. But this is what James was speaking of in this Scripture.

For a clearer picture of this, read my book; "Three Days of Darkness and the Coming Flood. This concludes my discourse on the number eight.

Chapter Nine
Number Nine

Many people are not prepared to hear what the number nine represents in the Scriptures. For the number nine represents the lost. One might ask how I arrive at this conclusion. It is not easy to see at first and it takes some study or someone to show how this number reveals those who are lost in the Scriptures.

To begin with, let's look at something Jesus said about the ninety-nine and the one lost sheep, which the shepherd goes after. Luke 15:1-7. Having listened to this parable over and over in Churches being interpreted as if the sheep which the shepherd goes after was being reinstated to the other ninety-nine for years, it dawned upon me that there was a need to explain this Scripture in its true light.

God has been gracious to me in allowing me to understand very deep truths, which He has placed inside His Word. I count it a blessing to be able to relate those things, which God has shown me over the years.

First of all; Luke 15:1-7 prepares us to understand this parable by letting us know the setting where Jesus is about to use this parable. Second, it explains to us that it IS a parable. Since it is a parable, it needs to be interpreted. But we must be careful how we interpret parables. We must not add to them mistakenly.

This is what I'm afraid has happened to this parable in every Church I have ever attended. They have added something, which is not within this parable at all. They have supposed that the lost sheep was brought back into the fold of the other ninety-nine sheep. This however, is not what Jesus said.

Looking carefully at Jesus' words, and applying the setting, we get a much different picture of what Jesus is actually doing with this parable. I would like you to notice that the publicans and sinners were drawing near to Jesus to listen to Him.

Jesus hears the Scribes and Pharisees murmuring about the fact that He received publicans and sinners. The Scriptures then tell us that Jesus addressed the parable of Luke 15:4-7 to these hard hearted Scribes and Pharisees.

Jesus asks them a question, which the answer to seems to elude those who read this parable. Jesus asks them, "What man among you, if he has a hundred sheep, and he lose one of them, does not leave the other ninety-nine in the wilderness and go after the one who is lost and when He finds it, he lays it on his shoulders rejoicing, and when he comes home, he calls together his friends and neighbors, saying unto them 'Rejoice with me, for I have found my sheep which was lost?"

I have yet to find a Pastor who understands that this is a legitimate question put to the Scribes and Pharisees. Jesus is asking them which man among them is like this man in the parable. The answer of course is that the only man who was like this shepherd in the parable is Himself (Jesus).

But we find something about this parable which is never addressed when it is spoken about in today's pulpits. Jesus says that the man in this parable takes the lost sheep and places it on his shoulders and goes home.

Jesus never says that the man takes the lost sheep back into the fold of the other ninety-nine sheep. No, what Jesus says, is that the man leaves the other ninety-nine in the wilderness unattended.

Now anyone with a lick of sense can tell you that sheep without a shepherd will become prey for wolves, coyotes, snakes, and any number of other things, which prey on them. This is exactly what Jesus is telling these self righteous Scribes and Pharisees which think that they are not in need of repentance.

Jesus is telling them that they will be left out in the wilderness as prey, while he goes after the one who realizes that he is lost. So what we see is, is that the ninety-nine sheep represent these Scribes and Pharisees and those who follow their false doctrines.

They believe that they are righteous and have no need of repentance. Therefore, they are lost and Jesus will leave them for the enemy to devour. This is why I say that the number nine shows those who are lost. Notice that the number nine in the number ninety-nine is twofold. This shows those who are twice dead or lost.

As with any truth in Scripture, it is expedient that we find a second witness to prove what we are saying. I therefore would like to draw your attention to an incident, which took place during Jesus' earthly ministry.

Scripture tells us that there were ten lepers who came to Jesus asking to be cleansed of their leprosy. Luke 17:12-19. Jesus tells them to go show themselves to the priests. On the way to do what He commanded them, they were cleansed.

Now all ten were cleansed, but only one of them turned back from going to the priests because he realized from where his healing had actually come from. The other nine continued on their way to the priests, thinking that they were healed because of their being obedient.

But it is the one who turned back realizing that it was Jesus' mercy that healed him, whom Jesus tells that his faith has made him whole or complete. This man was the only one who was saved and will end up in Heaven.

The other nine will be eternally lost. This is because they gave themselves the glory and not God. It was the one who turned back who realized that Jesus was more than a man. He came back and worshipped Him. He realized that Jesus was the long awaited Son of God who would come to save them.

So once again, we see that the number nine is associated with those who are eternally lost. Now unless I miss my guess, Judas who betrayed Jesus should also be the ninth Apostle called into the ministry. Let's check this out.

Okay, I was right. I first looked up Matthew 10:2, which names the twelve wondering what number Judas was mentioned, but he is mentioned last. But I knew I was onto something and did not let this get in my way. Judas is mentioned in Psalm 109:6-8. He betrayed a just man (Jesus) and had to be replaced.

I next went to the Old Testament figuring that the twelve patriarchs would be the next best place to look. Sure enough, when I looked them up, the tribe of Dan was the seventh tribe mentioned in the blessings of Jacob. Genesis 49:16-18.

Not to be daunted, I continued to look up Dan further in the Scriptures. Genesis 35:23-26 shows by the Spirit of God, that Dan is the ninth tribe mentioned. Why is this important? Because according to their birth order, Dan was the fifth son born to Jacob by Bilhah, Sarah's handmaid. Why then is he mentioned in ninth place, by the Spirit of God, when showing Jacob's sons?

It is important because of a Scripture, which shows that the tribe of Dan would be where the betrayer of Jesus would come from. This scripture is used by most in conjunction with anti-Christ, but it is actually speaking of Judas.

"Dan shall be a serpent by the way (Jesus is the way and Judas was by Him in His ministry), an adder (asp) in the path (the way); he (Satan) shall bite the horse heels so that his rider anti-Christ who rides a white horse in Revelation 6:2, shall fall backwards, (die) and yet recover (a false resurrection). The next verse says that either this adder has waited for Thy Salvation oh Lord (in order to kill him), or that this is Jesus speaking about waiting for His Father to declare Him righteous and resurrect Him. Not having a complete understanding of the Hebrew here, I cannot say for sure which is correct at this time, but I believe it is speaking of Satan lying in wait. Genesis 49:17-18.

But once again, the number nine is associated with the eternally lost. Many think Judas was forgiven. I beg to differ. If this were true, why does Jesus say, "It were better for that man, if he were never born?" Mark 14:18-21.

No, Judas was lost and was of the tribe of Dan, which was the ninth tribe of the thirteen tribes of Israel. Most forget that there were in fact thirteen tribes of Israel because the tribe of Levi was given no land of their own, but had to live on land the other twelve tribes provided.

Now I would like to propose something here which may offend some of my readers at first. But it came into my mind and I feel it deserves to at least be examined as a possibility.

When the Apostles realized that they were only eleven, after Jesus was betrayed by Judas, they drew lots to replace Judas' position of Apostle. It has always been a heated debate as to who was the actual Apostle who replaced Judas?

Some believe it was the one who won the lot when the Apostles cast the lot (rolled the dice) for Judas' position. This would be Mattathias. Some believe it was Paul the Apostle who replaced Judas.

I would like to propose something, which I believe is the truth behind the man whom Jesus actually chose to replace Judas. I would like you to notice that Mattathias was chosen by casting the lot.

This is also the action performed by those who took Jesus' tunic. They took something that did not belong to them by letting the lot make the choice. What this implies to me, is that Mattathias took something that did not belong to him by the same method.

God allowed this to take place, because Mattathias was also of the tribe of Dan. I believe that it will be one of his ancestors who will become anti-Christ. Now I know this will upset some of my readers because it appears that I am saying that Mattathias was evil.

Well, he may have been. But whether or not he was evil or not, I believe his ancestor will end up being the anti-Christ. This because I believe the above Scriptures, which speak of the asp in the way, refers to anti-Christ.

This shows us that Mattathias probably was an instrument of Satan during the early Church history. I cannot prove it, but Scripture seems to bear this up as a truth since he was chosen by lot (chance) and not by Divine decree.

No, it was Paul who was chosen by Divine decree. He was against Christ Jesus at first, and then repented. Now I have heard a false doctrine going around that says that Paul was a false Apostle. But it was not Paul who lacked a visitation from Jesus, but rather Mattathias.

Paul spent the rest of his life defending Jesus as Messiah. There is no mention of Mattathias doing this in the Scriptures. I would also like to bring a Scripture to mind for those out there who hold that Paul was a false Prophet or false Apostle.

2nd Peter 3:15-16, was written by Peter the Apostle. Peter calls Paul a beloved brother. I ask you, was Peter deceived? No, he knew Paul was a true believer, which is why he calls him a beloved brother.

No, I believe it was Mattathias who was not a true believer and who would become the father of those who were anti-Christ. These would give a good showing as if they were true followers of Christ, but in reality, they were only hypocrites, which means actors.

I believe this is why Mattathias, was chosen by lot. God allowed it to happen to show us that Mattathias was not a true Apostle. He was the next generation of vipers that would betray not only Jesus, but the Church as well.

Leaving this for you to decide for yourselves if I am on the right track or not in my above writings, I would like to continue by showing another Scripture which verifies that the number nine has to do with those who are lost.

In the Old Testament, God instructed those counting the cattle and sheep (pictures of types of men), to count the tenth animal (tithe) as holy unto the Lord whether good or bad. In other words, one out of ten was holy unto the Lord. The other nine were lost, or unholy.

So once again, we see the number nine associated with those who are lost. This is a pattern, which goes throughout the Scriptures. I would like to propose something else, which may or may not be accurate, but deserves mentioning.

In the story of Dinah, who was sister to Reuben and Simeon, she becomes a prominent figure in a story which tells us that Reuben and Simeon kill all the relatives of the leader of the people whose son wanted to marry Dinah after lying (having sex) with her.

Since Dan was the ninth tribe (spiritually as mentioned above), we may assume that it was not a tribe, which could be trusted. (I have already shown that both Judas, and the anti-Christ, both came and will come out of it).

So where does Dinah fit in? She would most likely be a picture of those who would replace those who were not worthy in the tribe of Dan. This may be why her name was Dinah as it is a form of the word Dan.

It may well be that by sleeping with the young man who wanted to marry her afterwards, she got pregnant. It would be her children, which would replace those within the tribe of Dan who God considered unworthy.

Of course this is merely conjecture on my part, but then why tell the story of Dinah's sexual encounter at all if it has nothing to do with the tribes of Israel? I believe it has very much to do with the tribes of Israel, and the fact that at least part of the tribe of Dan was unworthy to be called Israel.

This is the whole reason why God felt it important to include this story of Dinah within the Scriptures. He wanted us to understand that He would replace the lost members of the tribe of Dan with Dinah's descendants.

I'm not asking you to believe this as truth, just as a possibility. As I said, it is conjecture on my part and I could be off in my thinking. Nonetheless, it is an interesting theory. I don't mind juxtaposing some things.

Since I have ventured into juxtaposing, I would like to note an interesting thought concerning our solar system which was until recently, composed of nine planets. Recently, the ninth planet, Pluto, was decreed not to be a planet at all.

So once again, we see that even in science, the number nine is associated with that which is lost. One of our planets was lost. We now have only eight planets, which is the number of new beginnings. Our solar system has had a new beginning of sorts with this new declaration of our ninth planet being lost.

I find it interesting how science keeps following Scripture without even knowing it. Their decisions keep lining up with Scripture and yet they still do not understand. Once again I quote Scripture; "Ever learning and yet unable to come to the knowledge of the truth."

It is as if the hand of God is forcing them to make these decisions which line up with His Word. They have much to learn, as do many of the readers of this book. But time permitting; I shall endeavor to teach them and you the reader.

Before I continue with more numbers and their explanations, I would like to juxtapose one last thing. It has to do with the day of Pentecost. It also involves the number nine as it was the ninth hour when the Apostles made their speech to the visiting Jews from many countries around the world.

Acts 2:13-15 states that it was the third hour of the day. But in our clocks we have, it would be nine O'clock in the morning. This is because we count the whole day as twenty four hours. So from the beginning of a new day, it was the ninth hour when the speech occurred. Many non-believers were saying that the Apostles were full of sweet wine and drunk. Some translations call it new wine, but this is a mistranslation on their part.

New wine is wine that has not had any sugar placed into it to speed up fermentation. So the proper translation would be sweet wine implying wine to which sugar has been added to speed up fermentation.

Nonetheless, it was nine O'clock in the morning. We see God use this time to divide those in the crowd who are lost, from those who are to be saved. We find that there are about three thousand souls added to the Church at this time. Acts 2:41.

The word about or approximately is used here because even though there were three thousand men saved that day, some had wives and children who were also saved along with them. Scripture usually counts only adult males during a census.

But since God's Word prophesies that about a third of Israel shall be saved and two thirds lost, we can assume that there were approximately twelve thousand people who heard the message that day, not counting females and children.

You might wonder why I say Scripture says that only a third of Israel shall be saved. Paul said, "Though the number of Israel be as the sand of the sea, yet only a remnant shall be saved. Romans 9:27 and Isaiah 10:22. Zechariah 13:8-9, states that God will bring a third part of Israel through the fire to purify them.

Now the time frame of Zechariah 13:8-9 is when the shepherd was smitten. This would refer to Jesus being crucified. Matthew 26:31 and Mark 14:27. Therefore, God's prophecy about one third of Israel being saved at this time would mean that on the day of Pentecost, only a third of those who heard were saved.

So what we find then, is that there were about twelve thousand souls in attendance on the day of Pentecost, but only three thousand actually believed the message. This was God fulfilling Zechariah's prophecy.

So once again, we see nine thousand lost, and three thousand saved. Now you will say that the math is wrong. One third would make the crowd only nine thousand. True enough. But with the women and children, who are not counted, it was more along the lines of twelve thousand, and around four thousand were saved and eight thousand lost. Therefore, one third of them were saved and two thirds lost.

So it was during the ninth hour of the day if we count using twenty-four hours as our basis, that there was a separation between the lost and the saved of Israel. So then as with other Scriptures showing nine means lost, this one joins it in verifying this fact.

Chapter Ten
Number Ten

We continue our learning with number ten. This number correlates to the law. While Scripture states that God gave Moses ten words, one must go into the literal Hebrew to understand that these ten words in the commandments of God, only spelled out seven laws which could be subdivided into ten laws.

Be that as it may, there are enough Scriptures to show us that the number ten represents the law as well. I would like to draw your attention to the story of David and Goliath. Why? Because there is a subtle hint at this fact in this story.

The story of David and Goliath finds the men of Israel camped out against the forces of the Philistines. A giant by the name of Goliath is daily challenging them to come and fight with him.

Scripture goes into great detail as to how Goliath looked. Scripture tells us that Goliath was six cubits and a span. But what has this to do with the law? We must bypass what today's leaders would have you believe about the height of Goliath. We must see clearly what God is showing us.

Goliath was exactly ten feet tall. Scripture tells us that his armor was all made of bronze. Bronze in the Scriptures represents the flesh of mankind, which always incurs God's judgment. This is what God is saying in this true to life story, which He allowed to transpire in man's history.

The fear of Goliath was one of certain death to any man who would dare to confront him in battle. Scripture tells us that it was through the fear of death that Satan held us captive. Hebrews 2:14-15.

This is what this story which is an actual event in history, is trying to tell us in figurative form. Goliath represents the law, which condemned us because it was good, but we were wicked. Our flesh is represented by Goliath' armor which is all bronze.

The bronze of the armor could not stop Goliath from being killed. The rock, which David slung into the skull of Goliath, represents Jesus our savior who killed the law's right to judge us in order to bring in grace. Jesus did this by taking us to the cross.

The law has no dominion over a dead person. So once we have died with Christ, the penalty has already been paid for the breaking of the laws of God. But first, we had to die through Christ. This doesn't negate God's right to use temporary judgments upon His children to chastise them if they sin.

Goliath represents the fact that the flesh had broken the law. This is why he is ten feet tall. He represents the law, which caused us fear because of our wickedness. We knew there was a penalty for breaking the law. That penalty was death (eternal separation from God).

This is why all of Israel's champions of war are afraid of Goliath. God placed this fear into them to give a picture of the fact that mankind in his own strength could never defeat their destination of death for their breaking of His laws.

But God sent a champion to aid mankind. He sent His only begotten Son who died in our place. There is a paradox in this story of David and Goliath. Jesus is shown by both David and Goliath.

David represents Jesus' earthly ministry and His total reliance on God. Goliath represents Jesus on the cross. For Jesus on the cross became sin, which is who Goliath represents. 2nd Corinthians 5:21. Once Goliath died, Israel's fear was overcome. This is what happened once Jesus paid our penalty. Our fear of eternal death for breaking the laws of God was done away with.

Goliath's death represents Jesus' death on the cross. This is why he had to be exactly ten feet tall. He represented the old covenant under the law being done away with, so that the new covenant in God's grace would become effective.

Goliath's armor being made of bronze, symbolized mankind's poor efforts to appease God's wrath against him by human effort; good works so to speak. But these could not do away with God's judgment, which He pronounced on man in the Garden of Eden. Only death could accomplish the payment required by our sin.

Jesus paid the penalty of death on our behalf. This gave us boldness to approach God. Likewise, when Goliath died, the Israelites gained boldness to take their inheritance, which was promised by God.

There is another story in the Word of God, which gives us much the same picture. It is the story of Hezekiah's illness, which was cured by placing a fig-based paste upon his boil, which was killing him.

We find this story in 2nd Kings 20:1-11. Hezekiah is told that the sun will retreat ten steps back up the steps as a sign that he will recover from certain death. God is symbolizing that the old covenant of the law, which is shown by the symbolism of the sun, will be done away with, by covering it with darkness (a symbol of death).

The old covenant of the Ten Commandments was indeed done away with by the death of Jesus. God could then bring in the covenant of grace, which meant the new covenant could last forever. It would be an eternal covenant.

This is the reason why God promised to add fifteen years to Hezekiah's life. The number fifteen symbolizes spiritual completion. But I will get to this number later to prove this to you. Notice that the medicine, which cured Hezekiah, was a fig based mixture. This is because the fig represents the law. Satan used the law to kill mankind, God used it to save us.

To prove that the fig represents the law, we must go back to the beginning of mankind's creation. We must visit the story of Adam and the woman. It is there that the fig is actually first used to represent the law of God.

Adam is told by God not to eat from the fruit of the knowledge of good and evil. What most do not understand if any, is that the knowledge of something means to become intimate with it. It means to in effect become one with it.

God was telling Adam not to become intimate with God and yet at the same time be intimate with evil. Evil then, is a spirit. It is alive. God's jealousy raged when Adam became intimate with another spirit. The penalty was death. He committed adultery.

But God knew that it was through deception and seduction that the spirit of lawlessness was able to cause mankind to become intimate with it. This is why God decided to have mercy on mankind and curse the spirit of lawlessness or rebellion.

God told Adam not to eat from a certain tree. Most, if not all, do not understand that the tree that Adam ate off of at this time, which caused us to die, was a fig tree. We can prove this because of the type of leaves that they sewed together to cover their nakedness which is a type or picture of sin.

Scripture tells us that Adam's and the woman's eyes were opened after they ate of the forbidden fig. Trust me when I tell you that Adam and the woman were not created blind. Scripture is telling us that the eyes of their understanding were opened. They realized that they were not like God. They realized that they were different from God in one crucial way. They, unlike God, were sinful!

Immediately, they sought a cure from the sentence, which God had pronounced upon them; the penalty of death. They sewed fig leaves together and covered their loins with them. This symbolized to them, that their family line would survive.

Adam knew that the only way to cure an infection was to use the very same source from which the infection came. This is why the antidote for a snakebite, is made from the very same type of snake that bit the person.

Revelation tells us that the healing of the nations will be done through leaves. Revelation 22:2. So then Adam knew this fact which is why he tried to use the leaves of the same tree from which he had become infected.

So it was a fig tree which God commanded Adam not to eat from. This is why the law, is symbolized by a fig in the story of Hezekiah's illness. The same law, which killed mankind, saved mankind. Jesus, who is the law or Word of God, killed us on the cross with Himself, because of our wicked nature.

But Jesus also saved us, by fulfilling the law of sin and death. He paid the price or penalty, which was decreed by God in the Garden of Eden. So then the law or Word of God both slew us and brought us back to life. But the second life was by an eternal promise, which cannot be done away with.

Although the law was done away with, yet it still lives within us. For the Spirit of Jesus lives within us. This spirit can only obey the laws of God by its very nature. So then what appeared to be the death of the law only became the strengthening of the law.

We as born again believers in Yeshua (Jesus) understand that our flesh is evil and avoid allowing Satan to use us in rebellious ways against God's laws. As we mature in our walk, our rebellious flesh is being done away with. Therefore, we do not sin as often as we used to. 1st Peter 4:3-4.

I'll leave this number and press on to the number eleven, which has its own Scriptures showing what it means. Hopefully by now, you are gaining some knowledge in how God has placed hidden messages within the confines of Scripture for us as His children to find and understand.

Chapter Eleven
Number Eleven VS. Number Twelve

It has been a real pleasure to study and find these hidden nuggets within the Word of God. I am constantly amazed at the intricacies of hidden mathematical codes within God's Word. This book touches only the surface. For there are many hidden messages, which would take a lifetime to understand should one actually get into fine detail where mathematics and Scripture are concerned.

Nonetheless, I present the number eleven for you to see a little more clearly within the Scriptures. This number is illusive and its meaning not so easily obtained without serious consideration.

Let us start with the twelve Apostles. Judas was one of the twelve, but was only chosen to betray Jesus. When he hung himself, it left a void, which had to be filled. Hence the number eleven is shown as meaning incomplete.

I say this, because Scripture tells us that another had to take Judas' place. Acts 1:20 and Psalms 69:25 and 109:8. Here we see that the number of the Apostles was lacking or incomplete. So eleven then, represents something incomplete.

Another example where this is shown, is the story of Joseph. He is the eleventh son of Jacob. The name Joseph means, God will grant me another son. This implies that eleven is an incomplete number.

Later, we see Joseph has a brother named Benjamin, which came from his mother Rachel. This shows that it wasn't until Benjamin was born, that there was a proper foundation to build Israel upon.

Genesis tells us in what order these children of Jacob were born unto him. Reuben, Simeon, Levi, Judah. These four were Leah's children. Dan and Naphtali. These were Rachel's handmaid Bilhah's children. This made six children.

Gad and Asher. These by Zilpah. Isachar, Zebulun, and Dinah a daughter. These by Leah. Joseph and Benjamin. These by Rachel. Since we are talking about sons and not daughters here, Joseph was number eleven. Later, God added Benjamin which made the twelfth.

So now we see that again, God implies in His Word, that the number eleven means incomplete. Something is missing. There needs to be added one more for a foundation to be secure.

Also, you will notice that Dinah actually was number eleven and she was taken by force by the neighboring son of a tribal leader. Afterwards, she is never talked about. It is as if she died. I find this to be very strange indeed.

In Joseph's dream concerning eleven stars that bow down to him, we find that he is actually saying that his youngest brother will also bow down to him. Benjamin will bow down to him along with his other ten brothers.

Yet when Joseph speaks about his second dream, he says eleven stars bowed down to him and also the sun and the moon. This Joseph spoke of his eleven brothers and his father Jacob and his wife.

The interesting thing here is that Joseph's mother had already died. So how is it that she would bow down to him? This seems to be a prophetic Scripture. It must be talking about the Church. I say this in reference to Jesus' statement about whom He considered His mother and His brethren. Mark 3:31-35, and Luke 8:20-21.

This would seem to be the only plausible explanation since Joseph's mother had already died by this time. I could be wrong so don't quote me on this one. Nonetheless, it needed to be mentioned. The only other possibility I can come up with, is that the mother and father in Joseph's dream were Adam and Eve, from whom all men came. But even turning this into the spiritual mind would equal the saved or the Church.

Now we can go on to the number twelve, which Scripture seems to imply means a good foundation. We find this in both the Old Testament Scriptures as well as the New Testament Scriptures.

First we shall look at the Old Testament and the founding of Israel. God chose Abram whose name He changed to Abraham. This means father of many nations. So Abram became Abraham and went forth to spread God's message of salvation throughout the earth.

Jacob has his name changed to Israel by God. He has twelve sons, which become the twelve tribes of Israel. These twelve tribes of Israel become the foundation of a nation, which will spread the fame of God.

These twelve tribes are only a foreshadowing of the twelve Apostles who will bring forth the foundation of what is to be known as the New Testament Church. They become the founding fathers of today's Church age. Ephesians 2:19-20.

Jesus is said to be the chief cornerstone of the foundation. Notice that while Judas was among the others, Jesus Himself made the actual twelfth Apostle. Once Jesus left the others, Paul joined them to make the twelfth Apostle.

I have already given my view about Mattathias being a false Apostle. So then there were always twelve Apostles even when Jesus was amongst us. The number twelve then, symbolizes foundation.

If we look at the number of hours in a day, we see there are twelve. John 11:9. There are likewise twelve hours in a night. These are the foundation of our twenty four-hour day. When Jesus was about to be crucified, He said He could call out to His Father and He would send twelve legions of angels. Matthew 26:52-53.

What Jesus was saying, was that He could ask His Father for total deliverance and still be able to set up His kingdom here on earth. But then, how would the promises of God be fulfilled that He had already promised?

So according to Jesus' own words, it would take twelve legions of angels to build a foundation on which He could set up a kingdom if He was to go that route instead of the cross. But Jesus was not about to make His Father out to be a liar. His Father had made the plan of salvation so Jesus would stick to it.

When speaking of Ishmael, God promised Abraham that He would make of him twelve princes. Genesis 17:20. Although God would bless Ishmael's children, He would not make an eternal covenant with them.

Hence we see the twelve tribes of Israel which are symbolic of the children of light, as opposed to the twelve tribes if Ishmael who would make up the children of darkness. Twenty-four tribes; half for light and half for darkness; God keeping a balance between good and evil.

God had promised Abram that his seed would be as the stars of the heavens. Genesis 15:5. The thing is, is that God also promised Abraham that his seed would also be as the sand of the sea. Genesis 22:17.

Now what most do not understand, is that these two blessings are total opposites of one another. The blessed seed would number as the stars of heaven. The cursed seed would number as the sand of the sea.

Both types of seed would come out of Abraham. When God speaks of sand in the Scriptures, it is always in a negative connotation. This we see in many Scriptures. So what we find here is that God is keeping a perfect balance of both good and evil.

Let's take a look at some of the Scriptures, which speak of sand. Jesus spoke of those who would build their house on sand and that their ruin would be great. Matthew 7:26-27.

Likewise, we see that the beast who comes up from the sea, is seen while John stands upon the sand of the sea. This is indicative that the beast will rule over the evil people of this present age. They are the sand.

When speaking of the righteous, they are likened to the stars of heaven. We find this to be true in Daniel 12:3. So then there is a difference between the two blessings that Abraham received from God.

Abraham would be blessed (multiplied) in two areas; good and evil. This keeping a balance upon the earth, until God had finished what He had planned for those He intended to call His own.

This is what the Chinese refer to as ying and yang. They are not far from the truth when they speak of this concept. It is many Christians who cannot see this is a true formula in the Scriptures. An exact amount of evil or darkness as there is light or good.

Maybe you can now see this is a truth, which has been ignored by today's leaders. Nonetheless, I think with the above Scriptures, you can now see the pattern is there in Scripture to those wise enough to see it.

All things are balanced. God allows evil to exist for a reason. He uses it much like a tool. He shapes the children of light using evil for a godly purpose. Many cannot accept this truth and blind themselves to this simple truth.

It is beyond me why they choose to ignore these truths in Scripture. God himself created Satan. He created Satan with a goal in mind. He would use Satan to sculpture out the Church. She would be sculptured by her trials and tribulations brought about because of Satan.

If this thought offends you, just look at a couple of Scriptures with me. Isaiah 45:7 and Amos 3:6. These Scriptures plainly tell us that it is the Lord who creates calamity. He does it with a pure purpose in mind. He does it to train His children!

Most Christians balk at this interpretation. However, God is not going to sugar coat His Word so that we can swallow His truths. He has told us all things openly. It is up to us to accept these truths or to ignore them simply because they appear to be unfair.

Trust me when I tell you that God is doing only what He knows He needs to do in order to train His children correctly concerning good and evil. He uses evil like a tool to train His children, just the same as He uses good. Good and evil are alike to God. They are just tools He can use to train His own.

Evil is everything that God is not. So in order to show His children what He is not, He shows them His opposite, which is evil. God is good, everything, which opposes God, is evil. Yet God created Satan basically to test men.

If this were not true, then there would have been no reason for Satan at all, because God knows all things before hand. So since God knew Satan would fail Him, why create him in the first place? The only logical conclusion one can come to, is that God created Satan for the purpose of purifying the Church through trials and tribulations.

Now this gets into something, which has been a very controversial subject in the Church world today; predestination. This subject has caused major rifts in today's church bodies. The thing of it is, is that it is not something hard to understand.

The problem is not so much that it is a hard subject. The thing of it is, is that it steps on the pride of mankind. This is what the true problem is when dealing with this subject. People want to take away God's sovereignty.

To those who don't believe in predestination, I would like to ask you a very important question. Who decided that you were to be born into this present world? Was it you, or your parents? I dare to say you had no choice in the matter.

This is true not only in this world, but also in the next world as well. It is God who chooses to have his next child. We as human beings cannot take that sovereign right away from God no matter how we choose to twist the Word of God.

Jesus said it best; "You have not chosen Me, but I have chosen you, and ordained you, that ye should go and bring forth fruit, and your fruit should remain: that whatever ye shall ask the Father in My name, He may give you."

One does not choose God; it is God who chooses whom He will. Paul uses this same argument when speaking of what God told Rebecca about Jacob and Esau before they were even born. Jacob have I loved, but Esau have I hated. Romans 9:11-14.

But Paul continues this argument by adding what God told Moses on the mountaintop; "I will have mercy on whom 'I' will have mercy..." Romans 9:15. And again, "Scripture says unto Pharaoh, 'Even for this purpose have I raised thee up, that I might show my power in thee, and that My name might be declared throughout all the earth."

In other words, God made Pharaoh to have him put to shame, so that God's glory could be shown, by the miracles He used to deliver His people the Hebrews. This should be plain enough for anyone. Romans 9:17-18.

But Paul gets even bolder about predestination. He tells us that it is useless to argue against it. He even warns of people (and there are many inside the Church walls), who will argue that predestination is a false doctrine. Romans 9:19-23.

The above Scriptures argue that men will say unto God, "But that is not fair, why then are you sending anyone to hell since it is You who made them ahead of time to go there? They are not resisting your will then, since you made them this way."

But Paul answers on God's behalf; "Who are you O man that replies against God? Shall that which is formed say unto Him who formed it, "Why hast Thou made me thus?" In other words, God has power to form one for destruction and another for salvation. It is His decision since He made you!

Now that we have a solid foundation in these Scriptures, we can look at a misused Scripture which people always use to argue that a person has a choice to come to God and be saved of their own free will.

2nd Peter 3:9, truthfully states that God is not willing that any should perish. However, this is not a complete statement by Peter. The complete statement made by Peter, is that God is long suffering towards us-ward (on behalf of His chosen), not willing that any (of His chosen) should perish.

In other words, God has long suffered with the wicked of the world and will suffer yet more with their evilness, until all who are to be saved, have come into the fold of His children. This is what this Scripture is saying. It is NOT saying that God's desire is for everyone to be saved!

Scripture does not fight against itself. Paul is very clear that God chooses those to be condemned and those who are to be saved

BEFORE they are even made. Therefore, the explanation of the above Scripture IS its proper interpretation.

But then you will ask, "Why then should we even go preaching the gospel if a person is already destined to be saved before they are even born?" The answer is because God commands it. He commands it in order to build the character of those being saved.

What do I mean? Let me explain it so even a child can understand. Jesus said it best. "Love your enemies, do good to those who persecute you." Why? So that when God judges them, they will have nothing evil that they can say against you when they are judged.

This is the same reason why God makes the rain and the sun, shine upon the just and the unjust. Matthew 5:43-48. God is good to all. He wants you to be like Him. Preaching the gospel builds our character because we don't know who will be saved, but we, like God, are offering something good to all.

On God's part, He already knows which ones He made to be His children. But on our part, we only know that our God is good. We present Him as a loving Father to all. Thus we are witnessing that we know our Father is good.

This builds our character in the sense that we get to see our Father in Heaven manifest His goodness as we preach. This causes us to rejoice in our Father's goodness, and to rejoice when we see it happen for others.

God Himself calls the preaching of the gospel foolishness. 1st Corinthians 1:21. So if God calls the preaching of the gospel foolishness, then we can be sure that it is in fact foolishness. Yet God chose to use this type of foolishness.

It is foolishness, because God has already decided which people He made for salvation and which He made for damnation. He could have just made them and then separated them Himself. But He chose to allow us to become fellow workers with Him in order that He might reward us later.

It is the same idea of a son who is watching his father build a house they are to live in. The son desires to help the father who knows that he doesn't really need his little five-year old son's help.

The father allows the child to hold the nail he is about to hammer into the wood. The father knows that the child is actually impeding the progress that he could be making as far as building the house, but the father wants the child to feel a part of the project, so he allows the child to help him.

This is the idea behind allowing us to preach the gospel. It makes us feel as though we actually had something to do with building the house. The truth of the matter, is that we are impeding God's progress. But God, who is good, wants us to feel important, so He allows us to help Him.

This is why God commands us to preach the gospel. It is not so that those who are going to be lost have a chance to repent. They will never repent because they were made to be lost. It is so we as His children can have a part in salvation or the building of His house, which we are a part of if in fact we are truly His children.

When Paul speaks of his salvation, he tells us that when the fullness of time came, God's Son was revealed in him. He doesn't say that he received God's Son. He tells us that God's Son was already present within him waiting to be revealed. Galations 1:15-16.

In other words, God has already placed His Son within each and every one of us who is to be saved. It is just a matter of time before God's Son is revealed within us. It is like a seed that has been planted which sprouts when it is time for it to sprout.

So if the seed of the Son of God is not present within you, you can never be saved. Who planted the seed of the Son of God within those who are to be saved? God did. He did it while He was forming you in your mother's womb. It sprouts or comes to fruition at different times in different people. It just depends on the soil in which God planted the seed. Nevertheless, it sprouts because it cannot be held back by any type of soil, no matter how hard it is. God's seed always lives.

It is only a matter of time before it manifests itself. Those who are to be saved will receive the Word of salvation. But there are many, who are not saved, who are within Church walls. They have been called, but are not chosen.

Herein lies another thing that needs to be addressed. God calls many people to many offices in this lifetime. They will fulfill their missions of good or evil as God has ordained for them. But just because God called a person, does not mean He has chosen them for salvation.

What? Where does this come from? Scripture. Jesus said, "Many are called, but few are chosen. Matthew 24:14. Let us consider those who came out of Egypt at the time of the Exodus. Most died in the wilderness. God was not pleased with most of them and so destroyed them en route. They were all called out of Egypt, but few were actually chosen by God for salvation.

This is a foreign idea in most if not all Churches today. Yet Scripture is plain on this matter. Jesus tells us that near the end of time, just before He returns, deceptions shall deceive everyone except the elect (the chosen).

It is the chosen who are the true children of God, not the called. God knows that it takes many people to run an organization. The organization is called the Church, while God's true children are called the elect or chosen ones.

So God has called many workers, but not all of the workers He called are actually saved. This is why many shall come to Him in that day saying, "Did not I work many miracles in your name and cast out many devils?" His answer will be, "Depart from me you workers of iniquity, I never knew you."

Many who are working very hard within the Church walls are not actually saved at all. They are not chosen by God. They have merely been called by Him to do a task, but they will not be saved. This is why

Peter tells us to make our calling and election, not just our calling, sure (steadfast or secure). 2nd Peter 1:10.

Peter is careful to add the word election. Why? Because Peter understood, many are called, but few are chosen. He is addressing his words to those who are BOTH called and chosen. Otherwise, he would have just said; make your calling sure (steadfast or secure).

This is where confusion sets in concerning the fact that the Church MUST go through the tribulation period in order to be purified by trials and tribulations. Most Churches believe in a pre-tribulation rapture. This is NOT Scriptural.

They use the fact that the word Church is missing from chapter four onwards to convince people that the Church has been taken off of the earth. This is NOT what Revelation is saying. It is merely acknowledging the fact that the word Church is a generic term which merely means those who have been called out by God to do a temporary service for His true children which are known in Scripture as the elect or chosen.

This is the reason why God removes the word Church from Scripture at this point. From here on out, God will deal only with the elect to purify them. But you will have to read my other book; "Three Days of Darkness and the Coming Flood," for a full explanation of this concept and why the Church (the elect, not the organization) MUST endure the tribulation period according to Scripture.

It would take rewriting the above book to explain it here. I am not about to go through that process twice. But it is available to those wise enough to actually search out the truth concerning the difference between the Church and the elect.

If I were to sum up the difference in a nutshell, it would be that the elect actually love God with a pure heart. They follow His leadership. The Church merely wants to be saved, but have no true love for God. They have a facsimile love or pseudo love. They don't follow His laws; they merely want the benefits of salvation.

If I might be so bold as to offer a way to tell if a person is saved or not, it would be to see if they follow God's Commandments because they want to as opposed to because they have to. They simply love to do the right thing. It is their new nature.

Another sign of salvation is that they understand the Word of God accurately. I have already shown that many Scriptures are being misused. This is surely not by the Spirit of God. But there are other evidences. Love being the chief evidence.

When I say love, I don't mean love like the world means love. I mean love like God means love. In other words, love corrects when it needs to correct. It doesn't accept bad behavior or sin. It rebukes and restrains it.

Chapter Twelve
Thirteen the Priestly Number

Okay, so we have arrived at a number that is very, very controversial. This number is so controversial, that hotels refuse to number their floors as thirteen. This is due to many superstitious beliefs concerning this number.

First I would like to place something into this book which I believe to be a true story. Not that you have to believe it, but it bares mentioning. It concerns the Friday the thirteenth superstition and where I believe it came from.

Many years ago, there was a certain warlock, which was known as a high priest of his craft. Now the town Church, which was a Catholic Church, decided to end the influence of this man upon the people.

They gathered a group together and went to the warlock's home. They proceeded to drag him out into the public square and after lashing him many times, burned him at the stake for his crime of witchcraft.

This occurred on what we now know as Friday the thirteenth. The witches in the area proclaimed this to be sacrilege and made it a high day or holiday to honor their fallen brother of witchcraft. This is why this day is known as an unlucky day. It became very unlucky for witches.

Why do I mention this? Because I believe that there are times in history when the Church has been instrumental in doing away with evil when society fails to act upon its influence. This would be one of those times.

While our weapons are not carnal, but spiritual, there is no shame in using physical punishment against evil. This is why God gave us rulers and judges. Since the saints are to judge the world (1st Corinthians 6:1-2), this means that they have need at times to judge in the here and now also.

This is part of our job. We cannot simply ignore the physical when judging things. This is why evil has become so prevalent in today's society. It is because the Church has ignored any type of physical intervention. This gives the wicked only more room to accomplish their ungodly goals.

I tire of hearing so many people who are ignorant of the role in the Church to intervene in matters of great significance in world history. It has been mostly the Catholic Church who has had the nerve to stand up against the tyranny of evil through physical means when necessary.

Many site that it was the Catholic Church who were responsible for the Crusades. Well, yes it was. But what they don't tell you is that if the Crusades had not happened when they did, the Muslim community would have been too well established for Israel to defeat them in 1948.

So God had a plan when He allowed the Catholic Church to invade the land of Israel. He had to make sure the Muslim community was weakened sufficiently for those who would liberate Israel in 1948.

Why then do Christians condemn the Catholic Church for doing something that in the long run had something to do with Israel retaking their land? Furthermore, it was the Catholic Church who was instrumental in stopping the black plague known as the Bubonic Plague in Europe.

Their leaders banned together and forced the European governments to separate all those who were diagnosed with Bubonic Plague. This forced separation, caused the plague to cease from being spread any further.

The Catholics reasoning was a sound reasoning. They used the Old Testament law for those who had leprosy. By separating those with the disease, they stopped the plague from killing everyone.

Why am I mentioning this? Because the Catholic Church had gotten a very bad rap for things which may well have been God ordained that's why! Those who know nothing of history have used some of these incidents to condemn what may well have been God inspired. Yet there are few who know enough about history to put things as they should actually be viewed.

This brings me to another incident, which I must put into this book. I cannot say for sure one way or another, but it needs to be addressed as a possibility. History says that the Catholic Church was killing people for their religious views at the time when Christopher Columbus sailed for the New World.

My question to you is this. Were you there? How do you know exactly what views the Catholic Church was challenging as heretical. You have been taught that it was merely a political move that the Catholic Church was motivated by.

But what if this was not the truth? What if the truth lay in some other area having to do with the defense of Scripture? I know what you have been taught. But this doesn't mean that you have the whole story.

Many Christians now believe that our present government is built upon the Masonic Temple religion. This religion claims to be Christian, but is demonically controlled. This may offend some who are Masonic members but oh well, the truth offends many.

What if the Catholic Church full well knew that this form of anarchy against the Scriptures was taking place and sought to put an end to it by destroying those who followed its teachings?

If this is the case, then the Catholic Church once again would be taking abuse for doing away with witchcraft. Why? Once again I reiterate; there is no shame in using physical force against something which is trying to destroy those whom God has placed in charge for our benefit.

Am I certain this is what happened during those times? No, but anyone with eyes to see can see the demonic symbolism involved with Washington D.C. Even our statue of liberty is based upon a heathen religion.

I am not against government, but I am against any form of government law that tries to pretend to honor our God and yet denies us the right to honor God by making laws that are contrary to God's laws. One of the foremost laws which does this is the legalization of abortion. It is not the government I am against, it is their laws against God.

There are many others, but I would be here at the typewriter for many more days if I get sidetracked into naming them all. My main concern is that some things, which have actually helped the Church, have been things that occurred using physical force against tyranny, which has sought to overrule her teachings.

The question will now arise as to whether or not I am Catholic. I do not consider myself any religion at all. I consider myself a born again child of God. The walls of denomination I refuse to be held by. God knows no such walls when it comes to His children.

I simply look at things with a logical mind. Not all physical, force is bad. This is something that the Church will have to learn during the tribulation period. For Scripture tells us that the anti-Christ will make war with the saints and overcome them. But why? Could it be because the Church should have been making war physically at certain times in history much like the Catholic Church did against the Muslims during the Crusades?

Could it be that this is the whole reason why anti-Christ will win the war against the Church during the tribulation? Simply because they have become so lax where the physical end of their reign is concerned, that when push came to shove, they were unable to hold their ground?

Yes, Scripture says that it will be granted to anti-Christ to win the war, but there has to be a reason. God is not simply handing His children over to slaughter for nothing. Apparently it is a Divine judgment against them.

Answer me this. If Christ will make physical war against anti-Christ at the end of the millennium, why is it that the Church feels it is so wrong to make war with the evil seed upon our planet now?

Preaching the gospel is indeed our first goal and should remain so, but there are times when there is a clear and present danger, which will not yield to the gospel and will take over if not physically stopped.

This is one area the Church will have to learn the hard way as she goes through the great tribulation.

I know most of you reading this book do not believe the Church will go through the tribulation period. Your belief is based upon misused Scriptures, which have been handed down from generation to generation. I refer you once again to my book; "Three Days of Darkness and the Coming Flood."

Let me share something, which I believe to be a truth shown in Scripture. You are free to accept it or not. When God took the Hebrews out of Egypt, he fought for them. Why? Because He knew that at that time, they did not know how to fight.

He led them into the wilderness and it was at that time that He taught them to fight. Later, we see them overthrowing many peoples. This because they now knew how to fight. Of course God helped them, but they still had to fight. These wars were physical although they also saw miracles along with the physical warfare.

This was supposed to be the pattern for the Church as well. She started out too weak physically to fight. She started out like the Hebrews needing God to fight her battles for her. But when she became strong enough with sufficient numbers, she was to use also physical warfare if necessary to hold back or destroy evil.

But somewhere along the way, they lost sense. Somehow, they now expected God to fight all their battles without having to get into the battle themselves. This is what happened in the days of Joshua.

The people of God refused to fight because they were afraid of the giants. Spiritually, our giants would be tyrannical governments or nations. Instead of doing physical battles, the Church decided to allow these types of governments to control them. Then they cried unto God for their misfortune.

Do you suppose that God will not use physical force to throw His opponents into hellfire? He knows they have and will continue to refuse His reign over their lives. He will put them into an eternal prison.

One might argue that we no longer need physical restraint because of the gospel. If this were so, we could do away with our present day jails, which house those who refuse to obey the law of the land. But this is a ludicrous thought.

And so it is with the government of the Church. She was supposed to guard the earth against the devil and his forces. She needed to have the opportunity to grow large enough to do this. So God in her early stages, performed many miracles to keep her safe.

However, when she finally grew strong enough to where she could fight for herself, she did not use wisdom to keep the battle going. Instead, she lost all understanding of her position of being responsible to fight against the devil on a physical level as well as a spiritual level.

She became lax and in her laxness, even began to wane as far as fighting evil in her midst spiritually. She allowed all manor of sin to exist within her. She threw out her court system, which used to exist within her, and gave complete control of her worship services to someone known as a Pastor.

She got rid of the other offices of the Church which God declares are for the edifying of the body. These offices are; Apostle, prophet, and evangelist. Since the teacher was below the Pastor in rank, they were allowed to remain in office.

So now, when a prophet speaks against something the Pastor is doing which is not in line with God's word, he is immediately labeled a lunatic within the walls of the Church and becomes an outcast. He is told that he is causing division within the body.

Well, yes he is. He is causing a division between right and wrong! He knows full well what he is doing and has received his mission straight from the throne of God. But just as they persecuted the prophets of the Old Testament Church, so the New Testament Church has gone the same way.

Since these other offices have been done away with (are not listened to when they speak), the body has become very sick. The Church has suffered much due to lack of true leadership. The Pastor has no checks and balances to steer him into the right direction once he leaves the correct path. He can instigate any laws he chooses without question. In effect he become a dictator. And I say, this is exactly what has happened within the walls of the Churches.

There are many things being done within the Church, which fly in the face of Scripture. Many false doctrines have invaded the Church due to lack of leadership. Without these other offices operating within Her walls, she is unprotected against these false doctrines which have infiltrated her.

This book is not about these things. Get my other books. They go into full detail about prophecy and how it has been polluted by today's leaders. They tell of many practices, which have been polluted from their original intent. They speak of misused Scriptures, which have caused many prayers to go unanswered by God. How can God answer Scripture, which has been misquoted?

God indeed answers prayers if they are prayed with correct understanding of our Divine contract of marriage. But if we misquote the contract, He is under no obligation to answer such perverted requests. But to fully understand what I mean get my other books which go into detail as to what Scriptures are being misused.

Enough of history, I will get back to what the number thirteen represents. Once again I tell you it refers to the priesthood. But you are in need of proof and I must show this through Scripture.

Most teachers of Scripture conveniently forget that there were thirteen tribes of Israel. This is mostly due to the fact that they forget Joseph had two tribes within him. These are Ephraim and Manasseh. These two tribes are simply called Joseph in many Scriptural references.

The thirteenth tribe is Levi. Although Levi is the third son born to Leah, Jacob's first wife. However, scripturally speaking, Levi is the thirteenth tribe. We know this, because when the land was divided, Levi was given its land according to the positioning of all the other tribes. Hence it got its land last.

This showing that it was spiritually, the thirteenth tribe. It could not receive its land until all the other tribes had been established. It was in fact the priestly tribe. This shows us that the number thirteen represents the priesthood.

When God changed the priesthood in the New Testament, it temporarily went to Judah. It combined the kingly, prophet, and priestly order in Christ Jesus until He died. So Judah became the thirteenth tribe spiritually at this time.

But upon the death of Jesus, His spiritual name was changed to Ephraim. See my book, "Three Days of Darkness and the Coming Flood.

Now, the thirteenth or priestly tribe is Ephraim or more accurately so you can understand, the Church. The Spiritual name of the Church became Ephraim upon Jesus' resurrection. Once again I refer you to the above book.

No, I do not subscribe to the theory that England is Manasseh and the U.S. is Ephraim. This is not Scriptural. However, that the Spiritual name of the Church is Ephraim IS documented in Scripure. One just has to be estute enough to find out where this is found.

So now, Ephraim or the Church, has become the thirteenth tribe. She is now the priestly tribe and it is her job to fulfill all priestly duties. These duties include taking care of the sick which includes healing them, feeding the widows, orphans, and strangers, feeding and shepherding God's children with proper Scripture without twisting God's word.

I'd like to interject something here. The tribe of Manasseh was divided upon entering the Promised Land. Half of it remained on one side of the Jordan River. There is a spiritual reason for this physical event.

Manasseh was one of the twins born to Joseph. The half tribe of Manasseh which stayed on the east side of the Jordan represents the leaving of the flesh behind by the other half. It is indicative of the circumcision of the flesh a believer goes through upon coming to Christ Jesus. Their flesh cannot make the journey into the Promised Land.

The flesh must be left behind in order to serve God properly. But, just as the half tribe which stayed on the east side of the Jordan built an alter unto God to remind the other side that they were a part of Israel, so it is with our flesh, which shall become part of our resurrection later. God shall not forget to include our flesh in the regeneration.

One final note on the tribes of Israel. Dan is not mentioned in the list of tribes of Israel in Revelation chapter seven. This is due to the fact that they are where the anti-Christ came out of and have given themselves over to completely follow him. See Genesis 49:17.

Thus we see that the tribe of Dan was basically born as a tool for God to judge Israel's sins. This did not make the tribe of Dan a righteous tribe; it merely meant that God would use this tribe to bring evil upon Israel for their rebellions against Him. It is akin to why God created the devil. He was created, to be used as a tool to punish wayward children of God to drive them back to Him.

One might think this wrong theology. Okay, let's reason this one out for a minute. God knows all things. God is good. He can do no evil. So why create something He knows He is going to have to send to hell?

Why not rather, just never create him? After all, God knows the future completely. The reason is that Satan was a necessity. God intended for him to be a teaching tool for those who would inherit eternal life from God.

Does this sound cruel on God's part to you who are reading this book? Let me explain a hard fact about those who are not going to make it into God's kingdom. They were never really alive. What? That's right.

God's Spirit is life itself. Something or someone, who does not have His Spirit within itself or themselves, does not have true life at all. All true life comes from God. This is why being born again is absolutely necessary, in order to see life. The spirit one has when he comes out of the womb the first time is only a facsimile of life. They are in essence dead, not really alive.

Only God's life is true life. Without God's life abiding in you, there is no life. He is the author of life. He always lived. Before Him was nothing alive. Any of His creations, which don't have His Divine life inside of them, are not truthfully alive. They have pseudo life.

It is like when man creates a computer or android. It can laugh and with the proper fluids, can even cry and excrete other bodily excretions. But is it alive? No, it doesn't have blood like a human. It is run by oil and other fluids.

It would have to go through the birthing process within a woman to be called a real human being. So if I built an android to be scrapped later, am I being cruel? No, the android is not even alive!

Now the android might complain if it had some kind of reasoning power, but that would not change the fact that it was not truthfully human and therefore expendable. This is the case with those who are not born again. They don't have true life within themselves. They are like the android and are expendable.

This is why God has no problem sending the devil and his angels as well as humans to hell. He is not sending something alive into eternal fire. He is sending a facsimile of life into hell, an android.

Now those who are born again, ARE alive. God would never in His goodness, send a living thing into eternal punishment. He is good not evil. But the devil was a necessary creation to be used as a tool to teach God's children. Much like computers are used today to teach us things.

The only difference is that God's teaching tools had to be much more sophisticated for us to fully understand what God wanted us to understand about good and evil. They had to be very much like us.

Otherwise, God could not have taught us as much as we needed to learn. But they were never really alive.

How could they have been? God's Spirit is eternal. It can never die! If at any time, they actually had God's Spirit abiding in them, they could never have died (been separated from God) because they would have been His actual children. God is a good Father; He would never abandon His children.

This is how we can be sure that once saved, always saved. God has not given us temporary life. He has given us ETERNAL life. This life comes straight from God. It by its very nature is eternal! God doesn't abandon His children. He does however, need to train them.

Chapter Thirteen

Number Fourteen the Church; Then Add Christ

Scripture never really comes right out with its truths. They are like fine metals and jewels, which must be pulled out of the earth, using various methods of refinement. Scripture uses this same analogy in Job 28.

With this thought in mind, let us continue applying our hearts to Wisdom. Wisdom is the principle thing. We must therefore apply ourselves to study in order to find Her out. She is more valuable than fine metals or jewels. She is life itself. Without Her, we are dead souls. God possessed Her in the beginning. Proverbs 8.

In the story of Samson, we find that he had seven locks of his hair, which had never been shorn or cut. These locks of his hair were where his strength abode. Now if we think logically, hair has nothing to do with a man's strength.

So then, what is actually being said in the story of Samson? We must look deeper. Samson had long hair like a woman. These seven locks of hair were a picture of the sevenfold Church.

To understand how the Church is sevenfold in detail, see my book, "Time for Truth. Too much is entailed in the explanation to write a good explanation here.

Samson then, becomes a picture of Christ, who has seven Spirits of life within Himself which are His strength. The locks of hair are representative of these seven Spirits of life. But we still must look deeper.

These seven locks of hair which represent seven spirits, are indicative of the sevenfold spirit of life which Christ would give to the Church, both Old and New Testament. Since the Testaments are divided into two sections, we can safely say, that Christ gave this sevenfold Spirit of life to two different generations of men.

One generation lived before His crucifixion. They had these seven Spirits of life placed upon them while they lived upon the earth. They had to wait until Christ died to actually have these spirits placed within them. So they lived their lives experiencing this Spirit without actually having it revealed within them.

New Testament saints, who received Yeshua (Jesus), had a better covenant. They could have the sevenfold Spirit revealed within themselves, while they yet were alive. So we see this sevenfold Spirit of life given twice. Once for each covenant.

So if we add the two sevens together, we get a completed picture of Old and New Testament Churches. Hence the number fourteen in the Scriptures represents the Church both Old and New Testament.

Returning to the story of Samson, we see that when he is about to die, he asks God to give him his strength back one last time in order to avenge his two eyes. This is a play on words. It is a picture of Christ receiving back His life in order to avenge His two witnesses (Old and New Testament saints, as eyes speak of witnessing in Scripture).

You see the eyes, are what a human being uses to witness any event. This includes Jesus' death and resurrection. Hence God uses a play on words here to show that Jesus would regain his strength or life, in order to avenge His two witnesses, the Old and New Testament saints. The hair of Samson growing back symbolizes the Churches' resurrection. Of course this means that the cutting off of the hair symbolized their death with Christ.

As I said earlier, we must study if we are to find these nuggets of Wisdom. They are not handed to us without struggle. They must be wrested from the pages of Scripture. This struggle helps us to mature.

Okay, so where else is the number fourteen shown to be a picture of the Church. It is shown in that the number of spiritual completion is the number fifteen. This would be adding Yeshua (Jesus) to complete the Church whose number is fourteen.

But why do I say that the number fifteen is the number of completion spiritually? For this we need to study just a little more. Let's see why the number fifteen represents spiritual completion according to Scriptures.

Genesis speaks of a worldwide flood, which occurred to destroy all flesh except Noah and his immediate family. This flood is said in Scripture to have risen fifteen cubits upwards. All the high hills were covered, and the mountains were covered as well.

Fifteen cubits would be approximately twenty-two and a half feet. The question is, was this fifteen cubits altogether? Or was it fifteen cubits above the highest mountain? I believe it was fifteen cubits above the highest mountains. So the water rose twenty-two and a half feet above the highest mountains.

Also, Scripture is clear to denote the fact that the waters prevailed upon the earth for one hundred and fifty days. Genesis 7:24. The base number of one hundred and fifty would once again, be fifteen. So we see the pattern again.

In other words, God's judgment was complete. It took fifteen cubits of water, and one hundred and fifty days for God's spiritual judgment to be complete. Hence we see the number fifteen equals completion here. But where else can we see this to be true?

If we look to Revelation 21:16, the measurements are given for the New Jerusalem. This is not a literal city. It is a vision given to John the Apostle. It represents the Church. John cannot view the Church as She really is. Christ alone can view her as She really is.

There is one verse, which proves this is not a literal city. Revelation 21:21. This verse lets us know that there is only one street. The word is not plural. No city fifteen miles square could ever have just one street. But this city only has one street.

Of course, this is a reference to Yeshua (Jesus). He is the only road or way to God. Revelation 22:2 likewise tells us there is only one street. The word is once again singular not plural. So this is NOT a literal city. It is a vision of Christ and the Church.

Some translations say the measurements of the city are twelve furloughs long. The more accurate translations say fifteen miles square. So once again, we see the number fifteen is the number of Spiritual completion as opposed to the number six, which is physical or earthly completion.

In another example, we see Hezekiah is given fifteen years added to his life. This would complete God's judgment on Hezekiah who turned away from God in his later years. He did not honor God with the extension of his life. Instead, he did much evil. He would have been better off to allow God to take his life from him fifteen years earlier while his life was still honoring God.

But God allowed Hezekiah's words about always following God throughout his life to be proven wrong. For if Hezekiah had always followed God as he claimed, why all of a sudden did he falter after being graciously given fifteen extra years? Thus God showed him what he truthfully was by allowing him the extra time to fill up his sins.

So we see God used the number fifteen in Hezekiah's life to completely show Hezekiah's spiritual wickedness or sin. Then, God judged it. There are other examples of the number fifteen showing spiritual completion, but these should suffice to get my point across.

Chapter Fourteen
Salvation and the Number Seventeen

We've come to a very controversial subject here. This number seventeen means salvation. It is actually a variation of the number eight, which means new beginnings. For it is by adding one plus seven, that we get the number eight. Many would balk at this type of what they call numerology.

However, there is just cause for adding these two numbers together. First, the number for the Church is seven as I have shown. The number for Christ is one since He is God. Therefore, since Christ and the Church are one, we must add them together. Thus we come out to eight which is the number of new beginnings.

However, this new beginning begins with salvation. So where do we see Scripture showing us that the number seventeen means salvation? This is the controversial part. For if I get into too much detail here, I would have to rewrite my first book, "Three Days of Darkness and the Coming Flood within this book. But writing it once should be sufficient.

Therefore, I will try to keep the evidence from getting out of hand in this book. "Three Days of Darkness and the Coming Flood," gives ample evidence. But let's continue learning.

Noah went into the ark on the seventeenth of the second month according to Scripture and was saved from the wrath of God or the flood. Genesis 7:11. This begins our foundation for believing this number means salvation.

The second Scripture I'd like to bring to your attention is that God made the covenant with Abram on the anniversary of the day Noah went into the ark. Both events happened on the feast of Passover. See Genesis 15:12. A horrible darkness engulfs Abram. This is the same darkness, which came over Egypt. It lasted for three days.

The proof is that when the Hebrews were about to leave Egypt, there were three days of horrible darkness, which came upon Egypt. See

Exodus 12:41. It was exactly four hundred and thirty years later on the anniversary of Abram's covenant, on the very same day that they left Egypt. This after the three days of darkness.

Now you may argue that Genesis says it was the second month not the first month, which brought the rains of Genesis to flood the earth, while Passover is clearly taught in Scripture to be on the first month.

This is true. However, if you look at the Scripture speaking of the flood, you will see that there is no calendar mentioned when God is speaking about the month of the flood. All you see is Noah's age. So then the second month refers not to the calendar year, but rather the second month from the day of Noah's birth. So Noah was born in February. The second month then would be April or Abib.

So these three events occurred on Aril or Abib seventeenth. So God's covenant of salvation occurred on the seventeenth of Abib or April. Are you seeing a pattern yet? But Scripture goes further to show us this number seventeen, equals salvation.

The above things were meant only as types and shadows of the true event. For Yeshua (Jesus) was sacrificed for our sins on the fourteenth of Abib or April, and rose on the seventeenth day. So salvation for mankind was accomplished on the seventeenth day of Abib or April.

One more quick reference, the Hebrews conquered Jericho and entered into the land of promise on the seventeenth day of Abib or April as well. They rested at Gilgal on the tenth, but actually had to go around the city of Jericho for seven days. Thus they actually conquered Jericho on the seventeenth day of Abib.

Now some may try to disprove the above section as being true, because God told the people to rest for three days. It is clear this is a reference to the Passover, which started on the fourteenth. Thus if you add seven days to fourteen, it doesn't come out to seventeen.

While this is true, we must also remember, that it was not the common people who were marching around the city of Jericho. It was adult males who were twenty years and above who had been inducted into the armed forces that did the marching. The common people were allowed to observe the three days of rest from the fourteenth until the seventeenth.

The pattern of salvation remains pure throughout Scripture. It always occurs on the seventeenth day of Abib or April. So shall it be when Yeshua returns for His people. But there I go heading into my previous book, "Three Days of Darkness and the Coming Flood," I shall not rewrite it again here.

Chapter Fifteen
Thirty the Number of Betrayal

Judas, what was his job according to Scripture? His job was to hand over Yeshua to those who hated Him. He fulfilled Genesis 3:15, "It (the enmity) will bruise Thy (Yeshua's) head, and he (Satan) shall bruise thy heel (the Church).

Why do I equate the heel in this Scripture with the Church, rather than the fact that Jesus' heels were bruised on the cross? Because the feet of a man in the Scriptures, always have to do with his wife.

We can see this in Ruth 3:4-9. In these passages we see that Ruth is instructed by Naomi her mother in law, to go to Boaz and uncover his feet. Now what kind of sense does this make if we take it literally?

Of course what Naomi was asking Ruth to do, was to go to Boaz and offer herself to him for a wife. She first had to uncover herself to show him what he would be getting. She then, was Boaz's feet, which Naomi asked her to uncover.

One must understand why the woman is considered a man's feet in the Scriptures. Typically, it was the woman who did business for the man, because she knew how to bargain. She knew how to get what she knew her husband was really after in any transaction.

The more beautiful the wife was, the better bargain she could finagle, out of any deal set before her. She would deal with men whom her husband couldn't make budge on a particular deal.

Proverbs 31:10-16 tells us this fact. "...she considers a field and buys it." So it is the woman who was often sent to do a man's business transaction. This because the man knew other men would fold under her charms.

Thus we see that to uncover a man's feet, is to uncover the man's woman. This is what Naomi had asked Ruth to do. She asked her to undress in front of Boaz to show him what he would be buying if he took her to be his wife.

Another Scripture, which shows this type of language usage, is when David is about to cut the hem of King Saul's coat off while he is sleeping. David refuses to kill Saul, even though Saul had come to kill him.

But how did David get the opportunity to get close to King Saul in the first place? Scripture tells us that King Saul went into the cave to cover his feet. This Scripture is in 1st Samuel 24:3.What this tells us is that King Saul went into the cave to have sexual intercourse with his wife or concubine.

Of course his men would stand guard outside the cave; they would never have intruded on the king's privacy while he was being intimate with a woman. But David and his men were inside this cave, and waited until the king and his woman fell asleep and proceeded to secretly cut off the hem of Saul's robe.

David could do this, because Saul of course had taken his robe off in order to have intercourse with his woman. The robe was of course nearby, but not on the king himself. So David would not have to worry about waking the king up while cutting the robe.

So here again, we see that to uncover one's feet, is to uncover a man's wife or woman concubine. Scripture is filled with this type of usage of idioms. It simply takes a little time to realize what Scripture is actually trying to tell us.

But back to Judas and his betrayal of Yeshua; Judas received thirty pieces of silver for his betrayal. Thus we see that the number thirty equates to betrayal. It is interesting to note that the Old Testament calendar has exactly thirty days per month to it.

Our New Testament has thirty-one. Why the difference? I believe it is because our betrayal of Yeshua as a whole race of human beings was forgiven at the cross. So God moved on mankind to change the calendar.

When looking at the book of Revelation however, we see a return to the Old Testament Calendar of thirty days per month. Why? I believe it is because at this particular time in history, many who claimed to be Christian will show their true colors of betrayal towards Yeshua by handing over Christians to anti-Christ to be slaughtered.

Once again, we see mathematics and Scripture are closely tied together. I wonder if the addition of one day each month may have something to do with the earth's rotation. Did the earth's rotation change after Yeshua rose from the dead?

Could it be that the earth's rotation slowed down at this critical time in history? This would make the days longer, as it would take more to make one rotation. Of course this is speculation, but something caused our calendar days to change.

It is quite possible that the earthquake, which occurred when Jesus rose, was worldwide. This, because all Old Testament saints in every part of the world rose from their graves at this time. They had been liberated from death's grip.

This indeed could have caused the earth's rotation to slow down enough to change the amount of days in our present calendar. Or is it that the earth sped up? This would have made them add a day as well to each month as their calendar days would be insufficient to match the seasons. They would then have to add a day to each month.

Either way, it is quite interesting to note that the calendar changed once Yeshua rose from the dead. There are many who claim this is due to heathen calendar practices. However, if this were the case, then Rome surely would have changed the world's calendar long before they officially changed it, seeing Israel was already under their control.

No, I believe the resurrection of Yeshua caused something major to change in the earth's rotation. It demanded a change in our present day calendar. This is quite evident. But just what occurred, has been obscured by history.

Zechariah 11:12, tells us the price at which Yeshua would be valued by Judas. Thirty pieces of silver. For this amount of money, Judas would betray Yeshua into the hands of sinful men. But Yeshua knew this well ahead of time. He simply allowed it to occur so Scripture could be fulfilled.

One thought I would like to share with my readers is that Satan had to fill up his sins in order to be judged by God. So he was allowed to enter into Judas to betray Yeshua. Upon entering into Judas, Satan could also be judged, because he now had a fleshly body. Judas hung himself; thereby Satan's judgment was carried out.

I could bring up something here, which could cause some controversy, but I would instead like you to refer to my other book; "Time for Truth. This would cover something concerning Satan's origin, which I believe is pertinent here. I leave this to your discretion.

I would however, like to bring up this one point here. Upon Satan entering into Judas (see Luke 22:3 and John 13:27), Satan took on human form; he was flesh. This allowed God to deal with him after the manner of fleshly creatures. But this was not the first time Satan had been a fleshly creature. Satan was created a fleshly creature when Yeshua first made him. But as I said, read the above book for an explanation of this truth. That being said, let's move on to other numbers and mathematics.

Chapter Sixteen
Thirty Nine vs. Forty

Here we have two numbers, which are complete opposites. Thirty-nine speaks of mercy, while forty, speaks of judgment. Forty can also mean testing, but it seems always to end with mankind failing the test. The only exception to this rule is when Jesus was tested or tempted in the wilderness for forty days and nights. He alone passed the test.

Deuteronomy 25:3 allows for judgment up to forty stripes of a brother in Christ. The stripes may not exceed this amount lest thy brother seem vile unto thee. But Christ was spared the fortieth stripe out of mercy.

Yeshua had already been beaten by the guards and His beard plucked and had received a crown of thorns upon His head. There was no need for the fortieth stripe. This lack of a fortieth stripe speaks volumes of just how much Yeshua had already suffered by the hands of the guards.

But there is something else which bears mentioning here. There are exactly thirty-nine categories in the medical profession of diseases. This shows that Yeshua covered all medical expenses for our healing. This is why Scripture states that by Yeshua's stripes we are healed.

The punishment, which was to go to us, fell upon Yeshua in every category including death. This should set us free from fear. For God placed every category of Divine retribution upon Yeshua's shoulders to bear for us. The chastisement (punishment) of our peace (with God) fell upon Him (Yeshua). Isaiah 53:5.

With thirty-nine stripes, God healed us of all our diseases, having mercy on us who did not deserve mercy. The fortieth blow to Yeshua was His death on the cross. He paid our penalty of death, which the Lord God pronounced in the Garden of Eden. Genesis 2:16-17.

It is written that God tested the Hebrews in the desert for forty years. Hebrews 3:9-10. Although these verses speak of the Hebrews tempting or testing God's patience, it becomes apparent that it was actually God testing them. Of course they failed the test which God had placed upon them for forty years.

Yeshua was tempted of the devil for forty days and nights. Each day of Yeshua's test is the equivalent of one of those forty years, which the Hebrews failed. This is why God did not consume them with His wrath. It was because He saw Yeshua's time in the future when He would pass this forty-day test. Thus God accepted Yeshua's future sacrifice for these stubborn people.

This is how God was able to bless these stubborn Hebrews while they wandered in the desert. Their clothes didn't wear out, neither did their foot swell. Deuteronomy 8:4. They were fed each and every day according to Yeshua's future forty-day trial, which He passed on their behalf. Yeshua suffered the hunger they deserved for their sins against God.

So now maybe you can see how God was able to put up with this stiff necked people for forty years. He was constantly being reminded of Yeshua's future sacrifice on their behalf. Without this sacrifice of Yehua's, they would all have perished at once. They had all failed this test.

For each of the days that Yeshua went hungry, God fed the Hebrews for a year. So after 40 days and nights without food, Yeshua was tested with one last test. The devil came and offered Him the world if He would bow down to worship him. Yeshua answered him with the Word of God. "It is written, thou shalt worship the Lord thy God, and Him alone shalt thou worship."

This was the final blow to Satan's ego and the final blow to this test. God began Yeshua's ministry. He went forth with power to heal. The devil still tried to tempt Yeshua, but failed each and every time. Yeshua had set his mind on going to the cross DAILY!

This should be our goal as well. We will not obtain this level of obedience in this body, but we should not let this deter us from daily seeking to die to ourselves and live for God. This is our reasonable service unto God for all He has done for us.

I could cite countless Scriptures on the number forty here, but these should suffice to show that it means testing or judgment. Judgment because mankind always fails the testing. This may not be true on individual tests, but overall, we have failed God miserably. Thank you Lord for your eternal mercy!

Chapter Seventeen
Fifty and the Holy Spirit

Mathematically speaking, half of one hundred is fifty. This seems to be the idea here for using fifty as the symbolic number for the Holy Spirit. You see, the Holy Spirit is the significant other portion of God after which the Church is patterned.

In the beginning God decided to create mankind in His image. In the image of God created He them; male and female created he (mankind) them. This is one Scripture, which has not been fully understood by many theologians.

For this Scripture makes a fantastic declaration concerning the image of God which has been for too long overlooked by today's leadership. The above Scripture is found in Genesis 1:27. It states that the image of God is male and female.

Now I know, that you have been taught otherwise by today's leadership. The truth is that they fear to accept this statement made by God. They fear that it conflicts with another Scripture, which tells us not to create any image of God either male or female.

But this is NOT the case at all. For Hebrews tells us that Yeshua is the exact representation (image) of God the Father. However, this leaves us without the image of the Holy Spirit, which God states He would create. Genesis 1:26.

Since God is speaking to someone else in this Scripture, it would behoove us to learn who this other person is. The other person is none other than the Holy Spirit or the other half of God.

We can see that the Church is the other half of the image of God, if we look closely. Revelation 4:5 tells us the Holy Spirit is sevenfold. Revelation 1:20, tells us the Church mirrors this image and is also sevenfold.

The Church is the bride of Christ. So then it follows that the Holy Spirit would be the bride of God the Father. Yeshua's Mother. Now if this seems blasphemous, take another look at Genesis 1:26 and find a better explanation if you can. Who is the Us?

Furthermore, in Genesis 2:24, a prophecy concerning Christ is given which says, "For this cause a man (singular not plural as if it applies to all men), shall leave His Father and Mother, and shall cleave to His wife and the two (Christ and His bride the Church) shall become one flesh."

This Scripture is repeated by Paul the Apostle in Ephesians 5:30-32. He also says that it refers to Christ and the Church. So then the one who left His Father and Mother is Christ. Now one may argue that Christ left His earthly Father and Mother to start His ministry, but that would be a misnomer.

For Joseph, Yeshua's earthly father had already died before Yeshua started His earthly ministry. So this is not what the Scripture is saying. It is speaking of the Holy Spirit being Christ's Mother.

This is the idea for using the number fifty to represent the Holy Spirit. It is saying that the Holy Spirit is the other half of God the Father. This is why it was fifty days after Yeshua's resurrection that the Spirit of God fell upon the Apostles.

I would like to settle something here once and for all. Many believe that the coming of the Holy Spirit at Pentecost was when the Church began. This is not true. For Scripture tells us that when Yeshua resurrected, He gave the commission to His Apostles to preach the gospel to every creature. This occurred, the same day He resurrected. See John 20:18-23. These Scriptures tell us it was on the same day Yeshua resurrected that the commission was given and the Holy Spirit entered His Apostles when He breathed upon them.

So what was the purpose of the day of Pentecost? Like any good mother in law, the Holy Spirit came to guide her new daughter in law into all truths concerning Her Son. She would teach her daughter in law how to properly worship Her Son. She would teach the Church how to be a good wife to her Husband Christ.

This is the reason why the Holy Spirit came on Pentecost. She came to teach the Church how to be a good wife. So then, we see two Holy Spirits in the Scriptures. One was given by Christ to His bride to give Her life. The second, came to guide Her (the Church) into all truth.

This should clear up a much debated subject. For many cannot figure out if it is the Father or Yeshua who actually gives the Holy Spirit. Yeshua's Holy Spirit indwells His Church, this Holy Spirit came out of Christ upon His death to indwell the Church, while the Father's Holy Spirit is sent to teach Yeshua's bride and abides upon not inside, Christ's Church.

It really is not that complicated. What makes it complicated is the lack of sound teaching within the Church. This only adds confusion unto the body. How I ache for Christ's bride to learn these things.

Maybe this is the first time you have heard this explained this way. Maybe now you can understand why it was necessary for the day of Pentecost to arrive in the first place even though Yeshua had already given His Holy Spirit to the Church. I hope this clears things up for you.

So fifty represents the other half of God or the Holy Spirit. Fifty is one half of one hundred. Mathematically, it represents the other half of God. I hope you can see this now. Our study of numbers is completed. But there is something I would like to show my readers, so I will add two more chapters to this book.

Chapter Eighteen
Metals within Scripture

This chapter will be very short. I will not go into too much detail except to explain what certain metals mean when you see them in Scripture. Let's start with the metal Gold. This metal speaks of kingship. We can see this to be true when God gives a vision to Nebuchadnezzar.

Daniel interprets this dream and states that the head of Gold represents the King. So gold represents kingship. Daniel 2:38. This is why the three wise men (astrologers) brought Yeshua gold. This shows they understood He was a king.

The next metal I'd like to reveal the meaning of is silver. Silver speaks of the title of prince, or son ship. This is why Scripture tells us about the woman who swept her house until she found her lost silver coin. This speaks of a soul, which has been redeemed by Christ. They have become a prince or son of God.

God uses this metal in a negative connotation when speaking of those who have known the Scriptures and yet have refused to follow God. He calls them reprobate (impure) silver. These should have been saved, but because of their rebellious hearts, God has rejected them. Jeremiah 6:30. They shall be rejected sons.

Next comes iron, which means strength. But this metal implies strength against God most of the time. It implies those who are mighty in sin. They reject God's dominion in their lives. We see that anti-Christ's Kingdom is made of iron in the statue, which King Nebuchadnezzar dreams about.

But this type of strength speaks of devils. It doesn't speak of human rebellion. It speaks of the devil and his angels. Of course mankind joins with anti-Christ which is why the statue shows the clay along with the iron. But it states the clay would not join with the iron. This implies that the flesh would not entirely be as evil as the iron.

Next is the metal copper. This speaks of the flesh. It by its nature is always sinful. So copper speaks of the sinful flesh. This is why God had the utensils of the tent of meeting made out of copper and why they had to be sprinkled with the blood of the sacrifices. This shows that we as humans in order to serve God, must be sprinkled with Christ's blood.

When viewing Scripture concerning copper, always understand that it is speaking of a sinful act. This will give you a clear understanding of what is being said. The only exception is when the blood has been applied to the copper.

This concludes my explanation of metals. I hope this has helped you. Now, I would like to add a bonus chapter, which I believe is something that the Church needs to hear. It is something which God, has put on my heart.

Chapter Nineteen
What's wrong with the Church

In today's hustle and bustle world, it is easy to become distracted where Church government is concerned. Some all-important things in leadership are being overlooked within the walls of their congregations. I should like to address these things.

This section will hopefully stir those in charge to rethink how they are running their congregations and cause them to begin to reorganize how things operate within their walls. This is God's desire for them.

The early Church government had many things which today's Church government is lacking. To begin with, the early Church had a meal every time they gathered together. Now some of the people were not observing this meal correctly and were spoken to by the Apostles not to be separating themselves in clicks against those who had nothing to bring to these meals.

But the meals were a big part of the Church service. The members brought food from home, and would share with those who had no food. The deacons (who by the way were waiters, not businessmen; see Acts 6:1 and Acts 6:2.)

The deacons served tables like a waiter. So we see that there was indeed a meal at every service. This meal gave the members of a local Church the time to speak and fellowship with each other. True fellowship is not simply going to a Church service and lifting your hands together in worship to God.

True fellowship means to have intimate time spent with your fellow Church members so you can know them intimately; their sorrows and joys. This meal was intended to give that chance to get to know your fellow believers personally.

At this meal, if one announced a need, God would have already placed another member of the congregation at that table with the answer to that need. For example, let's say that someone at the table needed their transportation fixed. God would have placed a person with the ability to fix said transportation.

This person could then use his God given talents to help the one in need of a handyman. Someone who needed help with physical labor like chopping weeds down or tilling their farm, could get help from a person whom God had placed at the table to meet that particular need.

Someone who needed spiritual advice dealing with a certain subject could be helped along by someone in the congregation who had already been through that same problem. This was the idea behind these meals. The needs could be met at once. There was none of this "I'll pray for you," business.

The needs were met at once. The family of God came to the aid immediately upon hearing of the problem. Those who were able to meet the need heard about it immediately at the dinner table.

Unfortunately, this element of the Church has been taken away. This disallowed the members to become intimate with each other. Now, the members of a congregation think fellowship means to come to Church and simply say a "God bless you."

This is an injustice, which needs to be corrected within the Church of the living God. But in order for this to happen, the meal aspect of the service needs to be reinstated. I wonder if there are any leaders who will heed this message.

These meals were a far cry from the meals of today's Church which are made for profit. These early Church meals were for the purpose of keeping the body of Christ close. A person's needs were met immediately. They did not have to suffer needlessly. The body was one unit. Everyone helped their fellow believer.

It was this early love within the Church, which caused many to come to Christ. But this aspect has been done away with in favor of "I'll pray for you," which needs no real commitment. It is easy to pray for someone, not so easy to go to their house and actually help them.

Do these words sting? Good, then they are doing their job. I tire of people reading the Scriptures and not understanding what God is telling them. These things were written for our example so we could follow these godly lifestyles. Do you need to be hit with a brick to understand these things?

My words are harsh, yes. But how much harsher will God be when He calls the leaders into account for not doing these things when they are plainly laid out in Scripture? I will not mince words or keep silent on these things, as this is what God has called me to speak out against.

This is one of the major reasons why God refuses to perform miracles like the days of the early Church. How can He bless a people who refuse to obey His ordinances? The pattern is clear enough so that anyone who really wants to see it can implement it.

God refuses to perform many miracles because those in leadership as well as the common people refuse to do their part in helping their fellow believers. If you can't show God you love Him through taking care of each other's needs, don't expect God to do His part of performing those things, which are beyond your abilities.

Once again, my words are harsh. But God's will be even harsher when He judges for the lack of implementation of His laws, which are not that difficult to follow. Let the congregation have their meal so they can truly fellowship.

Let them hear about each other's needs over a meal and let those in the congregation who have the power to meet those needs meet them. Stop taking this right away from your congregation to use their God given skills to heal the body of Christ.

If you will at least do this, then God will begin to do what He used to do in the early Churches. He will begin to manifest His miracle working power. But He will not do it until He sees His people loving each other properly, which means meeting each other's needs that they have the power to meet.

Once this occurs, He will meet those needs, which we as humans are unable to meet. There will be miracles like the Church has not seen since its early years. Unfortunately, I know that few will listen. It would be too much trouble for them to implement this meal they would have to change how their service is run.

To these leaders I say; then don't expect to see many miracles in your Church body. God is tired of useless I love yous without the backing of it by proof, which means to actually meet needs when you are able.

God will again move in Churches who heed this message. Until then, don't expect God to manifest His power. He tires of useless words within the Church walls. He wants to see some action backing your words.

There are many things I could say in this chapter, but this is all I care to write about at this time. If the leaders of the Church can understand this message, it will greatly change the Church of the living God into a genuine instrument which will actually see the power of God move as it should have been moving all along.

It is the refusal to implement what God has clearly laid out in Scripture which has tied God's hands. How can He bless what He has not ordained? Running Church services as you desire instead of how God ordained them has caused God much grief as He watches His children's needs go unmet due to poor management by His leaders.

My desire is for you to grow in maturity. This is the reason for the writing of this book. May God grant you ears to hear and eyes to see. This is my prayer for those who represent God through Christ Yeshua.

www.ingramcontent.com/pod-product-compliance
Lightning Source LLC
LaVergne TN
LVHW090413210926
848480LV00002B/551
9798224704446